SAANICH ETHNOBOTANY

Culturally Important Plants of the W̱SÁNEĆ People

Nancy J. Turner and Richard J. Hebda

Featuring the botanical knowledge of Saanich elders
Elsie Claxton (Tsawout), Dave Elliott Sr (Tsartlip),
Christopher Paul (Tsartlip) and Violet Williams (Pauquachin).

Linguistic transcriptions by Timothy Montler, John Elliott Sr
and Dr Earl Claxton Sr.

ROYAL BC MUSEUM
Victoria, Canada

Reprinted 2019.

Published by the Royal BC Museum, 675 Belleville Street, Victoria, British Columbia, V8W 9W2, Canada.

Library and Archives Canada Cataloguing in Publication

Turner, Nancy J., 1947–
    Saanich ethnobotany: culturally important plants of the W̱SÁNEĆ people / Nancy J. Turner and Richard J. Hebda.

"Featuring the botanical knowledge of Saanich elders Elsie Claxton (Tsawout), Dave Elliott Sr (Tsartlip), Christopher Paul (Tsartlip) and Violet Williams (Pauquachin)".
Includes bibliographical references and index.
ISBN 978-0-7726-6577-5

    1. Ethnobotany – British Columbia. 2. Indians of North America – Ethnobotany – British Columbia. 3. Plants, Useful – British Columbia. I. Hebda, Richard J. (Richard Joseph), 1950– II. Royal BC Museum III. Title. IV. Title: Culturally important plants of the W̱SÁNEĆ people.

QK98.4 C36 T87 2012   581.6'308997940711   C2012-980150-X

For all the children of the W̱SÁNEĆ Nation.

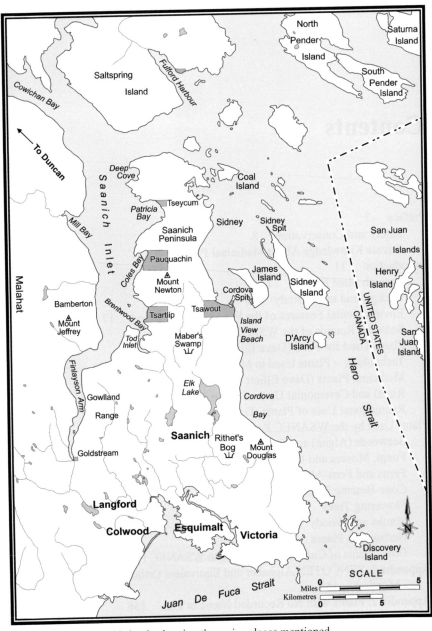

The area covered by this book, showing the major places mentioned.

Preface

This book was developed over a long period of time. We began putting it together after several years of ethnobotanical collaboration with W̱SÁNEĆ elders Elsie Claxton and Violet Williams in the mid 1980s to early 1990s. Together, we conceived the idea of including the teachings of these two knowledgeable elders. Both Elsie and Violet were concerned that much of what they had learned about plants through their lifetimes would be lost to their children and grandchildren, and to all W̱SÁNEĆ children, when they passed away. We talked about this idea with them, and with Salishan linguist, Dr Timothy Montler, who helped us to write the SENĆOŦEN names accurately using the International Phonetic Alphabet (which we have translated into a more practical alphabet). The dream of including Elsie's and Violet's rich botanical knowledge in a book was shared by family members, especially Dr Earl Claxton Sr, Belinda Claxton and Earl Claxton Jr (Elsie's son, daughter and grandson).

As the idea developed, we discussed it with Dr John Elliott, Earl's friend, colleague and cousin, and John's sister, Linda Elliott. We thought it would be a good idea to include the information from their father, Dave Elliott Sr, who, in 1980, wrote a booklet called "How the W̱SÁNEĆ People used Plants" for the Saanich Indian School Board. Dave Elliott Sr had developed a writing system that he used in this booklet, and John Elliott and Earl Claxton used this system to teach the SENĆOŦEN language at the ȽÁUWELṈEW̱ School. We decided this should be the writing system we would use in this plant book as well. After Violet's death in 1993, we discussed the book concept with her family and after careful thought they agreed to support its publication. We have also included information, published previously, provided by Tsartlip (W̱JOȽEȽP) cultural specialist Christopher Paul in interviews with Nancy Turner in 1968 and 1969 (see Turner and Bell 1971). We are deeply grateful to all these elders for their wisdom, knowledge, kindness and generosity. We are also grateful to their family members for supporting this project.

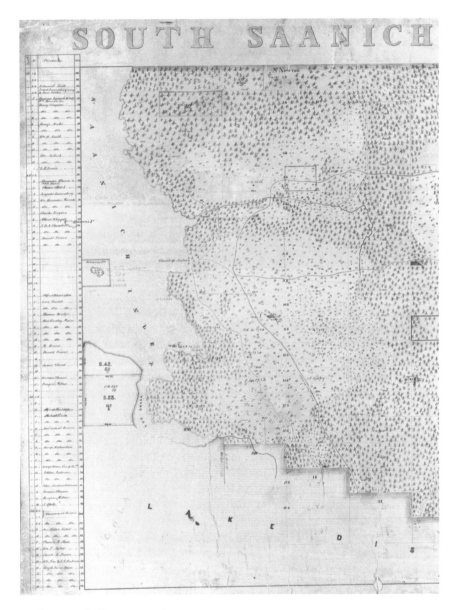

Safety and Conservation

It is very important to learn about plants before attempting to use them for any purpose. Before cutting off the branch of a tree to use as a skewer or part of a grill to roast food, make sure the wood is not poisonous or would make the food taste bad. This may seem like common sense, but every plant has its own characteristics. Almost all of them are useful to humans in some way, and all have important roles to play in the web of life, but these are not always apparent at first glance.

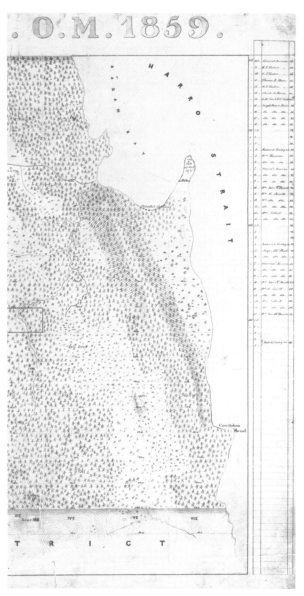

An early survey map of the centre portion of the Saanich Peninsula showing key landmarks and the distribution of forests (conifer tree symbols) and more open areas (other symbols). (BC Archives CM/B645.)

Many plants contain harmful chemicals that can cause serious injury if consumed. One example is Meadow Death-camas (*Zigadenus venenosus*). Its bulbs and leaves are violently poisonous, even deadly, whereas blue camas (*Camassia* spp.) bulbs are edible and nourishing if properly cooked and processed. The two plants grow together, so it is important to be able to tell them apart.

With medicines, it is extremely important to collect the right plant and know exactly how to prepare it before attempting to use it as a medicine. Many medicines can be harmful or poisonous if taken when wrongly prepared or in the wrong dosage. Even one kind of plant can vary in the concentration of medicinal ingredients it contains, and therefore in its power and strength for healing. Traditionally trained plant specialists will tell you that all plants have their own spirit or life force, and that any plant you wish to harvest and use must be treated with great respect. Talking to the plant and asking for its help and permission to harvest it may seem strange to those coming from an urban, western perspective, but this is standard practice for many indigenous users of plants and other resources.

Safety for yourself and others is a major concern, but so are the conservation, health and wellbeing of the plants and their habitats. Many useful plants are becoming rare and hard to find. It is very important to look after these plants so that they do not disappear. For the sake of the plants, you should find out how common they are before collecting them. Some plants can be grown easily in your garden or even in pots.

General caution: Never experiment with medicines or foods without consultation or guidance from a knowledgeable plant specialist or a qualified health practitioner.

Private Knowledge About Medicinal Plants

We recognize the private and sacred nature of medicinal plants and other preparations used ceremonially or ritually by the W̱SÁNEĆ (Saanich) People. It is not our intent to violate the privacy of the elders who shared their knowledge with us. Their concerns were to ensure that the identity of the plants that were so important to them is kept alive in the W̱SÁNEĆ culture and taught to the children, to be passed on to the future generations. Dr Earl Claxton Sr, our advisor for developing this book, noted that, for the medicinal plants to be truly effective, they need to have the special words or incantations pronounced. It is these words accompanying the harvesting and preparation of the plants that give them supernatural healing powers. Therefore, Earl explained, just including the information that a certain plant was used as a medicine, even saying how much was used and how it is prepared, does not violate the privacy of the family-held knowledge of its use. The accompanying incantations are available only to those entitled to hold them.

Introduction

There was no money long ago, but it's nice.... Lots of food, lots of clams, lots of wild berries all over. Five dollars would buy you sugar, coffee, tea and flour. That's all anybody needed. Everything else they got from the land.
– Elsie Claxton, 1996.

Ours was an abundant land. Our forests, meadows, creek sides, marshes and sea shores offered many plants for our use. – Dave Elliott Sr, 1980.

Plants have always been important to the W̱SÁNEĆ people, as to all peoples of the world. Trees, shrubs and other kinds of plants, including seaweeds, are major sources of food, materials and medicines for humans, and provide the backdrop and environmental texture for all cultural activities. Plant names and terms relating to plant harvesting and processing activities are a significant component of languages. The study of the many interrelationships between people and plants is called ethnobotany. Ethnobotany helps all of us to understand the importance of plants for human health and wellbeing, and the role of plants in human culture and language.

Saanich Plant Experts

For the W̱SÁNEĆ people of the Saanich Peninsula and adjacent islands, their traditions reflect a close relationship with the plants of their territory. In our original study we have recorded in as much detail as possible the plant knowledge of two contemporary elders of the Saanich Peninsula near Victoria – Violet Williams and Elsie Claxton. The compilation of their ethnobotanical knowledge is supplemented with information obtained from other individuals, including Christopher Paul of the Tsartlip Band, interviewed by Nancy Turner and others in the late 1960s and early 1970s (see Turner and Bell 1971), and Mary Thomas of Esquimalt, Violet Williams's elder sister, who was consulted from time to time by Violet during the course of our work, and was interviewed with Violet and Elsie by Nancy Turner during one session in 1990. To this information we added details provided

The Saanich Peninsula seen from Malahat Ridge. Mount Newton is in the centre. Pauquachin is at its base on the left and Tsartlip is at the right edge of the shoreline. Tsawout lies on the opposite shore behind Mount Newton. The Gulf Islands and San Juan Islands can be seen in the distance.

by Dave Elliott Sr in his report, "How the Saanich People Used Plants" (1980). Some of the information Dave included was drawn from Christopher Paul's earlier information (cited in Turner and Bell 1971). Our book is not intended to be a complete ethnobotanical study of the W̱SÁNEĆ peoples. It focuses on the knowledge of a few individuals, but we assume much of the information to be more widely known among the W̱SÁNEĆ community, both past and present.

Elsie Claxton (XEȾXOȾELWET)

Elsie Claxton was born and raised at Tsawout (East Saanich), and always spoke SENĆOŦEN. Her father and his mother and uncles all lived there. Elsie's mother moved from West Saanich to East Saanich when she got married. Elsie's mother's mother was from Chemainus. Her mother's father's name was apparently Michelle. Elsie Claxton's brothers – Phillip, Elliot, Sandy, Marshall and Albert Pelkey – have all passed away. When Elsie was between 8 and 14 years old, her mother worked at the cannery at Sidney, where she and her co-workers canned clams, crab, fish, apples, and plums. Elsie recalled picking large quantities of Wild Blackberries (SḰELÁLNEW̱, *Rubus ursinus*), Soapberries (SX̱ÁSEM, *Shepherdia canadensis*), Salal berries (DAḴE, *Gaultheria shallon*), and Red Huckleberries (QEḰĆES, *Vaccinium parvifolium*) when she was a girl. Elsie's work as a knowledgeable elder and teacher of W̱SÁNEĆ ways, culture and language, is being carried on by her children and grandchildren.

Left to right: Elsie
Claxton, Violet Williams
and Richard Hebda at
Goldstream.

Violet Williams (ŁIḴELWET)

Violet Williams was from BOḰEĆEN (Pauquachin or Coles Bay) and so were her father and his parents. Her mother and mother's family were from Westholme, near Duncan, and Violet and her sister Mary lived there as girls. Her great grandmother was from Duncan and great grandfather from Mill Bay. Vi noted that many of the people who used to live at Mill Bay moved across to Coles Bay some years ago because it was too far away to come over all the time to attend the Indian dances. In the old days they had dancing in long houses at Patricia Bay, East Saanich (Tsawout), Tsartlip and Deep Cove.

Vi's first language was Quw'utsun' (Cowichan), a dialect of Halq'emeylem. When she came down to Pauquachin, she was taught to speak SENĆOŦEN by Elsie Claxton and Elsie's grandmother. Elsie and Vi were friends for many years, and in our study on plants, they worked together, in many cases confirming each other's recollections of names and uses of plants. (They were also related – Elsie's husband was Violet Williams' father's cousin.) During our interviews, Elsie Claxton preferred to speak in her own language and have Violet Williams translate her words into English for us. Both women had extensive experience with and knowledge of wild plants. Much of this knowledge consisted of recollections of plant use from their childhood and young adulthood, and of information passed down to them by their parents and grandparents. Both women used many wild plants as sources of food and medicine right up until the time of their deaths.

Vi's elder sister, Mary Thomas, a member of the Esquimalt Band, also participated in the study. We interviewed her once, and Violet checked names and uses of plants with her.

Left to right: Elsie Claxton, Mary Thomas and Violet Williams.

Dave Elliott Sr

Dave Elliott was born in 1900. He was raised at W̱JOȽEȽP (Tsartlip) and lived there all his life. His mother was Saanich and his father was an Englishman who worked as a logger during the construction of the Island Highway and at the cannery at Sidney. His mother also worked at the cannery. When Dave was still a boy, his father was killed in an accident at the cannery, and he was raised from that point by his mother and her family.

Even as a small child, he travelled around with his mother and his aunt, to the San Juan Islands and many places around Saanich territory, learning about fishing, berry picking and the traditional lifestyle. He also spent time in his early years with his JOMEḴ, his great-grandfather, and his great-grandmother who lived at W̱SEYKEM, Patricia Bay. He always had a keen interest in traditional W̱SÁNEĆ ways, and in educating young people about their importance. Dave Elliott's knowledge and teachings in this book come from both published and unpublished sources (as cited in References). He had a particular expertise in the ocean, the tides, fishing and sea life. Today, Dave Elliott's work in maintaining W̱SÁNEĆ culture, language and lifeways and in educating the Saanich children has been taken on by his children. For this book, John Elliott and Linda Elliott have been key consultants and supporters.

Christopher Paul Sr

Christopher Stephen Paul was born in 1893 on a 60-acre farm at the corner of Stelly's X Road and West Saanich Road at W̱JOȽEȽP (Tsartlip), the son of Thomas Paul of W̱JOȽEȽP and Annie Michel of Quamichan. His brother was the famous wrestler, Chief Thunderbird. Christopher was raised in a traditional way – he participated in spiritual training and education, in traditional methods of food procurement and navigation, and he learned traditional ecological knowledge. His grandfather, Tommy Paul, worked with Homer Barnett (1955), who thought highly of him. Christopher also learned about plants from his mother's mother,

Christopher Paul.

who was Quw'utsun' (Cowichan), and from other elders in the community. He fished and hunted, and maintained his own garden, where he grew vegetables, strawberries and even camas.

A kind and gentle man, Christopher Paul was generous to students such as Nancy Turner, who learned from him as an undergraduate in botany in 1968 and 1969, and Marguerite Babcock, an anthropology student doing a field-studies project in 1967. He taught the Saanich language and culture to many others, and at the age of 73, he started teaching at the Tsartlip Indian School. He also taught both Saanich and Quw'utsun' languages at the University of Victoria, Camosun College and St Ann's Academy, as well as at W̱JOȽEȽP. He passed away in the late 1970s. His family members still reside in the same location where he lived. His knowledge and teachings in this book are drawn from Turner and Bell (1971), based on Nancy Turner's interviews with him and on literature sources.

Our interviews with Violet Williams and Elsie Claxton in this study extended over many sessions from 1987 to 1991, and, for Elsie, until the time of her death in the spring of 2000. Their plant knowledge was compiled mainly by Nancy Turner and Richard Hebda. The linguistic transcriptions of plant names and botanical terminology were provided by Timothy Montler, a linguist who has been studying the SENĆOŦEN and Klallam languages since the late 1970s.

During most of the interview sessions with Elsie and Vi, we brought out plant samples for discussion and naming. Violet and Elsie would discuss the names and uses of the plants that they recognized. We elicited specific information concerning the preparation of the plant for its use. In some cases we mentioned the SENĆOŦEN name as reported by previous consultants to help Violet and Elsie remember the plant. A collection of voucher specimens has been made and deposited at the herbarium of the Royal British Columbia Museum.

Background to the Study

This study is part of ongoing ethnobotanical research among indigenous peoples in British Columbia and neighbouring areas, beginning with such well known ethnographers as Franz Boas and James Teit, and continuing to the present with various ethnobotanists and linguists working together with aboriginal elders. Ethnobotanical knowledge is recognized as an important component of cultural education programs.

Among Coast Salish peoples, aside from various published sources, there are a number of unpublished documents and compilations of field notes pertaining to ethnobotanical knowledge. These include records on traditional medicines by Dr David Rollins, and on ethnobotanical details by Randy Bouchard and Dr Dorothy I.D. Kennedy, David Rozen, Diamond Jenness, Dr C.F. Newcombe and Dr Thomas Hess. Violet Williams and Elsie Claxton's knowledge of tree-bark medicines has been published separately (Turner and Hebda 1990).

The elders who shared their knowledge with us in this book have recognized more than 120 species of plants and provided Saanich names for most of them. Although the ongoing loss of language and traditional knowledge among aboriginal cultures must be recognized, it is encouraging to note that significant ethnobotanical knowledge has been retained into the 21st century, and with growing interest in cultural heritage and languages among younger generations of aboriginal peoples, particularly in school programs, the prospects for the passing of this important knowledge on to the following generations are good.

We hope this information will be learned and used for years to come by the W̱SÁNEĆ people, especially the children and grandchildren of the elders whose knowledge is featured here.

Wooded and open landscapes of the Saanich Peninsula. Mount Newton is in the background with communities of Brentwood Bay and Tsartlip on the lowlands in front of it.

A coastal Douglas-fir forest on Saanich Peninsula.

Environmental Features of the W̱SÁNEĆ Homeland

W̱SÁNEĆ territory, on southeastern Vancouver Island, falls within a diverse land-scape of rocky shoreline, sandy beaches, estuarine flats and coastal bluffs, with wooded, sometimes rugged hills and occasional open meadowlands. Most of the lower areas fall within the dry subzone of the Coastal Douglas-fir forest zone (Meidinger and Pojar 1991).

Most of the W̱SÁNEĆ lands are forested, or at least they were in the past. But many open dry habitats occur, especially on the Saanich Peninsula and around and within the city of Victoria. Wetland and shoreline communities occur along the coast and scattered lakes and wetlands occupy depressions throughout.

Douglas-fir (*Pseudotsuga menziesii*) is the most common tree in upland stands of coastal Douglas-fir forests. More than a century of logging and dis-turbance have removed the giant trees that once towered over the land. Now, second-growth stands of Douglas-fir cover much of the territory. In less disturbed settings, Grand Fir (*Abies grandis*) grows together with Douglas-fir. Western Redcedar (*Thuja plicata*) is abundant, especially in moist sites, whereas dry sites favour Arbutus (*Arbutus menziesii*) and Garry Oak (*Quercus garryana*). Other frequently encountered deciduous trees include Red Alder (*Alnus rubra*), Bitter Cherry (*Prunus emarginata*), Bigleaf Maple (*Acer macrophyllum*), willows (*Salix* spp.) and Western Flowering Dogwood (*Cornus nuttallii*). Copses of Trembling Aspen (*Populus tremuloides*) occur scattered in the drier sections of this zone,

Saanich Inlet with Malahat Ridge above the far shore and Douglas-fir forests on the slopes.

and Tall Black Cottonwood (*Populus balsamifera* subsp. *trichocarpa*) usually inhabits moist disturbed settings and edges of lakes, such as Elk Lake. Four other conifers – Western Hemlock (*Tsuga heterophylla*), Sitka Spruce (*Picea sitchensis*), Lodgepole Pine (*Pinus contorta*) and Western Yew (*Taxus brevifolia*) – are also found in W̱SÁNEĆ territory, though infrequently. All of these woody species are known to and used by W̱SÁNEĆ people.

The western part of the W̱SÁNEĆ territory borders another forest zone, the moist Coastal Western Hemlock vegetation zone. The W̱SÁNEĆ people sometimes visited sites in this zone and used the plants occurring there. Western Hemlock (*Tsuga heterophylla)* and Western Redcedar dominate this zone, leaving few openings in the forest. But Douglas-fir remains an abundant species in the dry parts. The driest wooded sites in the territory support a community dominated by Garry Oak, Arbutus and Douglas-fir, sometimes mixed with Lodgepole Pine. Oceanspray (*Holodiscus discolor*) is an especially common shrub in this open woodland, and there are numerous other kinds of shrubs. Moderately dry sites feature almost pure stands of Douglas-fir, sheltering a shrub layer of Salal (*Gaultheria shallon*) and several other shrub species, especially Oregon-grape (*Mahonia aquifolium* and *M. nervosa*).

Sites with moderate moisture and rich soils support a mixed canopy with Douglas-fir, Grand Fir, Western Redcedar, Bigleaf Maple and Western Flowering Dogwood. Many different shrubby and herbaceous (non-woody) species occupy openings as well: Thimbleberry (*Rubus parviflorus*), wild roses (*Rosa* spp.), Trail-

Garry Oak-camas ecosystems on Mount Tolmie.

ing Blackberry (*Rubus ursinus*), Oceanspray and Mock-orange (*Philadelphus lewisii*), to name a few. Wet situations with rich soils especially support Western Redcedar, Red Alder and Pacific Crabapple (*Malus fusca*). Several fruit-bearing shrubs, including Salmonberry (*Rubus spectabilis*), Indian Plum (*Oemleria cerasiformis*), Red-osier Dogwood (*Cornus stolonifera*) and Red Elderberry (*Sambucus racemosa*) favour moist rich soils. Skunk-cabbage (*Lysichiton americanus*) is particularly characteristic of the herb layer in wet sites, but several fern species and relatives, including Lady Fern (*Athyrium filix-femina*), Spiny Wood Fern (*Dryopteris expansa*), Sword Fern (*Polystichum munitum*) and Giant Horsetail (*Equisetum telmateia*) are common. The ground is usually covered in a lush carpet of herbs and mosses.

Natural meadows and open rocky hilltops are home to many plant species and resources. Numerous shrubs form thickets in the transition between wooded plant communities to open ones. Common shrubs here include Oceanspray, Snowberry (*Symphoricarpos albus*) and Tall Oregon-grape (*Mahonia aquifolium*). Thickets give way to grass-and-herb-dominated meadows, or grassy moss-covered rocky "balds". On deeper soils camas species (*Camassia quamash* and *C. leichtlinii*) combine with Indian Celery (*Lomatium nudicaule*), Western Buttercup (*Ranunculus occidentalis*) and diverse wildflowers. Originally these meadows had few grasses, but with the coming of Europeans many introduced alien grass species became established. Some of these, such as Sweet Vernal Grass (*Anthoxanthum odoratum*), have taken over from the native species. On the balds, scrubby oaks

Garry Oak woodland on a rocky knoll.

Camas and Western Buttercup meadow.

and shrubs seek moisture in deeper soil pockets in the rocky surface, while mosses, lichens and wildflowers (annual and perennial) occupy dry sites with little soil.

There are many other specialized habitats in W̱SÁNEĆ territory, though some were greatly reduced by European settlers in the latter part of the 1800s, who converted portions to agricultural and residential lands. W̱SÁNEĆ elders recalled one significant marsh, known locally as Maber's Swamp, for its rich, diverse resources. Settlers drained this swamp without permission from or consultation with the W̱SÁNEĆ people.

> The local Saanich swampy ground was seen to be a nuisance to the new arrivals to this land. So, consequently a huge, deep ditch was dug to drain the water away, without consideration to those who would be affected. Five little streams that passed through the village of W̱JOȽEȽP [Tsartlip] on the west side of the Saanich Peninsula and brought life to the bay had now gone dry…. When the immigrant people drained the swamp…, Cecelia Elliott [1865–1933] who was a Saanich elder at the time, cried and said, "I wonder if these people know what they are doing?"… She knew that this was the end of the place where the Saanich people gathered the willow bark for their reef nets and cedar bark for their anchor lines, baskets and their ceremonial dress…. Medicines, herbs, berries, ducks … swamp reeds for travel mats and house linings, cedar for canoes and … the boards [that] covered our homes…. All this, she could see and she could see the end of a beautiful way of life disappearing before her eyes. She knew that troubled times lay ahead for her people. (Claxton and Elliott 1994.)

Tim Montler also recalls Elsie Claxton expressing her puzzlement about how stupid the settlers were to ruin such a place of bounty.

Black Cottonwoods at
Maber's swamp in winter.

The extent of Maber's Swamp is shown on the early surveyors' maps. Parts of it were acidic and boggy, historically supporting a large Lodgepole Pine bog forest in a central zone surrounded by swamp or sedge marsh. Around such wet areas Hardhack (*Spiraea douglasii*) and willow communities would have occurred as they do today in smaller wet places. Within the stands of Lodgepole Pine there once were extensive thickets of Labrador-tea (*Ledum groenlandicum*), Bog Cranberry (*Vaccinium oxycoccos*) and possibly other bog species typical for the region (Roemer 1972). Later, after the bog was destroyed, the W̱SÁNEĆ people resorted to visiting Rithet's Bog in Royal Oak as a source for Bog Cranberries and Labrador-tea.

Other notable wooded assemblages in moist settings included mixed coniferous forests of Western Redcedar and Red Alder preceded by successional stages including Sitka Spruce and Western Hemlock. On the southeastern extreme of the Saanich Peninsula swamps consisted mainly of stands of cottonwood, Trembling Aspen, Slough Sedge (*Carex obnupta*) and willow (Roemer 1972).

Today only Rithet's Bog in Saanich remains. Although the edges of the wetland have reverted from agriculture to willow thickets and marsh, the central Lodgepole Pine-Labrador-tea forest struggles to survive, and the peat mosses (*Sphagnum* spp.) and Bog Cranberry are almost gone.

Shoreline plant communities are of three types: saltmarshes, estuaries and beaches. Scattered patches of saltmarsh occur along the coast of southeast Van-

Shoreline habitat at Island View Beach, south of Tsawout.

couver Island. These mainly consist of communities of Saltgrass (*Distichilis spicata*) and Glasswort (*Salicornia virginica*) in the mid intertidal zone and diverse tidal meadows dominated by herbaceous species and grasses in the upper intertidal zone. In the middle and lower intertidal and subtidal zones there are other species important to the W̱SÁNEĆ, mainly Common Eel-grass (*Zostera marina*), with its sweet-tasting rhizomes, and Bull Kelp (*Nereocystis luetkeana*), a brown alga, whose long, tough stems were used to make fishing lines.

Extensive beach communities occur especially along the east side of the Saanich Peninsula, including a large area in the Tsawout Reserve. Grasses and wildflowers grow in profusion in this habitat. An important grass species for the W̱SÁNEĆ is Dune Wild-rye Grass (*Leymus mollis*). Also found along the upper beach are patches of Pacific Silverweed (*Potentilla egedii*), whose roots were eaten, and Indian Celery (*Lomatium nudicaule*), whose seeds were widely used for spiritual purposes and in medicine.

Today, many of the indigenous plant species and communities have been replaced by ecosystems with introduced weeds, such as Scotch Broom (*Cytisus scoparius*), Himalayan Blackberry (*Rubus discolor*) and Orchard Grass (*Dactylis glomerata*). The W̱SÁNEĆ people have learned to utilize some introduced species, like Dandelion (*Taraxacum officinale*) and docks (*Rumex* spp.), but others they have not recognized to have any particular utility.

As well as the W̱SÁNEĆ people, many different birds and animals depend on these plants and habitats for food and shelter. Long ago, Roosevelt Elk, Grey Wolves, American Black Bears and Cougars, as well as Columbian Black-tailed Deer, were common in W̱SÁNEĆ territory, but most of these were hunted out by the last part of the 1800s. Today you can still find deer, but elders like Elsie

Elsie Claxton (left)
and Violet Williams on
Cordova spit at Tsawout.

Claxton maintained that the deer of recent years are not as healthy and do not taste as good as they used to. Cougars live in the Gowlland Range of hills above Saanich Inlet and sometimes enter urban areas such as Sidney. Bears visit rarely and elk have not been seen for decades.

Seasonal Rounds of the W̱SÁNEĆ People

The language of the people is deeply connected to the natural world, its plants and animals, and the timing of natural events. SENĆOŦEN, or Saanich, is a dialect of Northern Straits Salish, a language in the Central Division of the Salishan Language Family. The relationship of Northern Straits Salish to other languages, both Salishan and non-Salishan, is described by Thompson and Kinkade (1990). Culturally, the W̱SÁNEĆ belong to the Northwest Coast Cultural Area. Ethnographic information on these groups may be obtained from Barnett (1955) and Suttles (1951, 1987, 1990).

The W̱SÁNEĆ, like other Northwest Coast peoples, had a traditional economy based on hunting, fishing and plant harvesting. Their permanent winter villages were established along the coast, in quiet bays, coves and river estuaries. During the harvesting season, families and small groups moved over wide areas to obtain particular resources, including berries, roots and other plant foods. Even in the last century, many W̱SÁNEĆ people spent time fishing and duck hunting at the Goldstream River estuary and along Finlayson Arm, and also travelled across Saanich Inlet to the Malahat territory to pick berries, including wild strawberries (*Fragaria* spp.), Red Huckleberries (*Vaccinium parvifolium*) and Salal berries (*Gaultheria shallon*), and to dig edible roots, such as Wild Onions (*Allium cernuum*) and "wild carrots" (probably Yampah, *Perideridia gairdneri*). People also camped for days, even weeks, on some of the islands to harvest camas bulbs and hunt. The Gulf Islands – Salt Spring, Saturna, Pender and others – were all favourite camping places for the W̱SÁNEĆ. The San Juan Islands in Washington are also a part of the W̱SÁNEĆ traditional territory. Earl Claxton remembers, as a young boy, going to camp on Henry Island, where people went to go reef-net fishing. Later, he went with his relatives to fish for cod in the deep waters next to the sheer cliffs on the island's north side.

The seasonal harvesting activities of the W̱SÁNEĆ are described in detail by Earl Claxton in a book called *The Saanich Year* (1993), based in large part on information recorded by Dave Elliott. Each family had its own special places and their own favourite resources and activities, but in general, the patterns of the seasonal round were on a 13-moon schedule, as drawn from Earl Claxton's book, with some notations from Diamond Jenness (1945).

Dr Earl Claxton Sr with a map of W̱SÁNEĆ territory.

SSIS,ET – The Elder Moon (December)

This is the winter moon, with short days and stormy, rainy weather. People mostly spent this time indoors, because travel on the ocean could be treacherous and unpredictable. People ventured out to gather fuel and to hunt the overwintering ducks and geese, to fish for a few cod and grilse (young salmon), and to collect clams and other shellfish. Most of their time, however, was spent in making netting from stored nettle stems, carving canoes, and making baskets and other items needed for the coming harvest season. Children listened to stories told by the elders, and spiritual and cultural activities such as the winter ceremonial dances were held in the longhouses. People mostly ate the dried fish and berries they had stored from the previous year.

NINENE – Moon of the Child (January)

This is the beginning of the Saanich year, as the days grow longer and the world starts its rebirth. There is some warmth in the air and a few sunny days, although there are still many days of cold, wet, windy weather. Families started to assemble their reef nets for the coming fishing season. They began to catch a few Spring (Chinook) Salmon and seals, and some ventured out to fish for halibut, but they still relied in large part on stored food. Grilse were good for catching at Tsartlip. This is the season when fawns were born, so people stopped hunting does at this time. The ceremonial dances and story-telling continued long into the nights.

WEXES – Moon of the Frog (February)

In this moon the frogs wake up and start their choruses each night, announcing the coming of spring. The earth warms up, and the rain is not as heavy. At this time people put their canoes back in the water and started travelling more widely. They fished for cod, grilse, Spring Salmon, halibut, and especially herring. And they harvested herring roe by placing cedar boughs in the water where the herring spawned; the earliest runs of spawning herring were in Fulford Harbour. People caught ducks, too, as they gathered to get the herring and their roe. They set up duck nets in the narrow passes between islands. The reef nets were assembled, and potential reef-net fishing sites were surveyed. The winter ceremonial dances were finished as people started to stay outdoors more.

PEXSISEN – Moon of Opening Hands, Blossoming Out Moon (March–April)

The plants open their leaves and the blossoms open. There is more sun, which is good to dry the food being harvested. People hunted and preserved the Brant Geese (XELXELJ) using float nets in their feeding grounds. They also harvested clams, oysters and mussels. Cedar trees were felled for canoes, and women started to strip the bark from cedar trees for weaving mats and clothing. Long ago, when people kept small, woolly dogs, this was the time of year when the dogs' wool started to shed, and it was collected and spun for blankets.

SXÁNEŁ – Bullhead Moon (April)

During this moon there is a big wind and the big bullheads (large-headed bottom fish) appear on the shore. This is also the time when the swallows arrive. The weather is generally good, but there can be sudden thunder-and-lightning storms. People spent much of their time on the water. The older women would spear the bullheads (SKA) from under the rocks. This was also the time to harvest Seaweed (*Porphyra abbottiae*). People stopped fishing for halibut because they spawn during this moon, but they could snare almost-full-grown grouse from the woods. Around this time the young shoots of Giant Horsetail, Cow-parsnip (*Heracleum maximum*), Salmonberry and Thimbleberry were harvested and eaten. They provided vitamin C and were welcomed as fresh greens.

PENÁWEN – Moon of the Camas Harvest (May)

People travelled to many locations to dig camas bulbs, "wild carrots" and other root vegetables. Particular families owned some of the better places for hunting, fishing and camas harvesting. People also gathered fresh seagull eggs, since the seagulls nested in the camas grounds on many of the islands. Purple and green sea urchins were also gathered to eat. Fishing for cod, Spring Salmon, grilse and deep-water halibut took place around this time.

ĆENŦEKI – Sockeye Moon (May–June)

This is the time to fish for Sockeye Salmon. The reef nets, which were being prepared earlier in the spring, were put in place and a ceremony was held as the first salmon of the season were caught. The Swainson's Thrushes are singing in full force, and the Salmonberries, strawberries and other berries are starting to ripen. It is said that the song of this thrush is what puts colour into the Salmonberries. The W̱SÁNEĆ people would trade their catch of Sockeye with other First Nations. Because they were able to catch salmon in the straits, they had them about a month earlier than the people who had to wait until the salmon started up the rivers to spawn. A special ceremony was held and a special song sung to pay tribute and thanks to the first salmon.

ĆENHENEN - Humpback Salmon Return to the Earth (June–July)
This is a hot, dry time, when the grasses and forests are dry and there is a danger of fire. People went far into their own and neighbouring territories to camp and fish for humpback salmon. There were large gatherings with extended families and neighbours from other villages, where people traded, courted and exchanged ideas and information. Feasts were held and the W̱SÁNEĆ people shared the wealth of their harvest with other people. Men hunted a few elk and deer around this time. These are prime berry-picking times, with the ripening of Trailing Blackberries, strawberries, Red Huckleberries, Wild Gooseberries (*Ribes divaricatum*) and others.

ĆENŦÁ,WEN – Coho Salmon Return to the Earth (August)
The first fall rains come to help swell the creeks and rivers and the Coho return to their spawning streams. People fished for Pacific Tomcod and Lingcod. The weather starts to cool and the deer-hunting season begins. Indian Celery seed is harvested and stored for its many ceremonial and medicinal uses, as well as its use for flavouring meat and fish. Salal berries, Saskatoon berries, some late-ripening Thimbleberries, Blackcaps, Stink Currants (*Ribes bracteosum*) and a number of other fruits are at their prime. They would be harvested and eaten fresh, as well as being dried in cakes for winter use. At this time, too, the men repaired their canoes and boats, or made new ones.

ĆENQOLEW̱ – Dog Salmon Return to Earth (September)
Rainy, windy fall weather sets in and the Dog (Chum) Salmon return to spawn in the swollen rivers. People hunt for deer and grouse on the land, and sea lions and seals on the sea. Cod fishing is at its prime. At Goldstream, people fish for Dog Salmon, the last of the salmon to be fished. They smoke the fish they catch to preserve it for winter. Cranberries and Bog Blueberries (*Vaccinium uliginosum*) would be harvested in the Langford area. Wild Crabapples (*Malus fusca*) and Hazelnuts (*Corylus cornuta*), too, would have been picked at this time. Some of the root vegetables, like Silverweed (*Potentilla egedii*) and Springbank Clover (*Trifolium wormskjoldii*), were dug. People held a first salmon ceremony to honour the Dog Salmon People, and attended potlatches in neighbouring communities. Women gathered clams, and made blankets and rush mats (from Tule, *Schoenoplectus acutus*, and Cattail, *Typha latifolia*). Wood was cut and stockpiled for winter fuel.

PEKELÁNEW̱ – Moon that Turns the Leaves White/Faded (October)

This is the end of the harvest season, the nights are longer and cooler and there is frost on the leaves and ground in higher country. People split logs for canoes, firewood and building materials. They hunted seals and sea lions in the San Juan Islands and made preparations to hunt deer and elk. The last berries and fruits would have been harvested at this time.

W̱ESELÁNEW̱ – Moon of the Shaker Leaves (October–November)

This is the beginning of the wintery weather, which kept people close to shore and close to their winter villages. Most food was preserved and stored away, and people fished only close to their homes. Some people moved west into the mountains to hunt elk, but only after the first snowfall so that the animals could be tracked down easily if they were wounded and wouldn't be wasted. People began their winter gatherings at this time.

SJELȻÁSEṈ – Moon of Putting your Paddle Away in the Bush (November–December)

This is a winter month, when people mostly stayed in their homes, sheltered from the winter storms. They sometimes ventured out in the night at low tides to dig clams. All the stored materials were taken out and worked: the women wove mats and capes and baskets, and the men made their fish nets, boxes, tools and fishing gear. They also worked with cedar, felled earlier, splitting it into boards for walls and roofing materials. Winter ceremonials began, and children once again got to hear the stories from the elders. People started to eat the food they had carefully stored during the harvest season.

Plants Used for Food

All or part of this and the following sections are from Dave Elliott Sr (1980) and, through him, from Christopher Paul Sr and other Saanich elders.

Some of our elders could list about 70 different plants used as sources of food. If all these plants were classified by a botanist, we would find that they represent about 30 different families of plants. These plants were important because they provided the necessary supplement to our diet of fish and meat.

The main plant foods were bulbs and berries. Many roots were also used. Each plant was collected at the time when it was best for our use. So, different plants were collected in different seasons. Cambium (inner bark of trees) was collected in the spring and eaten fresh or dried. Our people looked for a tall tree with few branches and cut at the base of the tree and pulled up to take a long strip of bark. We were always careful not to take too much, so the tree would not die.

Women were responsible for most collecting and preparing of plant foods. They made and used a great variety of containers for collecting the foods. Most often they carried baskets on their backs or over the shoulder secured with a tumpline to leave their hands free. Roots and bulbs were dug with a strong pointed stick. Cambium was scraped off with a stick.

Some plant foods were eaten raw but most were cooked. There were three methods commonly used for cooking plant foods. Some plants were roasted over open coals of a hot fire. Some were boiled in watertight containers; red-hot rocks would be put into containers to make the water boil. Some plants were cooked in steaming pits dug in the ground. This is called TWÁS in our language. The pit was lined with red-hot rocks, water poured over them, then the food was wrapped in large leaves and placed in the pit.

Many of these plant foods were preserved and stored for winter use. Bulbs and roots were steamed, then air-dried and hung up in Cattail (*Typha latifolia*) bags. Fleshy berries could be dried like raisins. Moderately juicy berries were mashed, some boiled, and then dried out in rectangular cakes. Cambium and seaweed were also dried in rectangular cakes.

Dried cakes of berries, seaweed and cambium were stored in cedar bent-wood boxes. In order to have these diet supplements through the winter great quantities of plant food had to be collected and prepared for storage. To eat them, dried cakes were either soaked or boiled in water. So, you see, we had a complete technology worked out for the gathering and use of plant foods. Each was col-lected in its season, some eaten fresh, and most of it preserved for winter use.

Technology – Plants Used to Make Things

We used about 50 different plants to make the things we needed. The plants were used in four basic ways: wood, fibre, dye and fire. There were also general house-hold uses for plants, such as pitch for patching and mosses for diapers. Wood products for the most part were made by the men. They made the houses, canoes, wooden boxes, bowls, masks, totems, bows, arrows and the smaller tools. They used many techniques to make these wooden items. Large planks were split off cedar logs, canoes were dug and moulded by using fire and boiling water, boxes were made by bending wood with steam.

Women worked with fibre materials to weave mats, baskets, bags, clothing, hats, blankets, twine and fish nets. They used shredded bark and roots, and the leaves and stems of grasses, marsh plants and some bushes. Most of the weaving was done by hand with some wooden implements to help in assembly. They used a one-bar loom to weave cedar-bark blankets and garments.

Dyes made from plants were usually extracted by boiling and the liquid simmered and used like a dip dye. Some barks contain high concentrations of tannin used for tanning and curing wood as well as hides. They were extracted by boiling as well.

Each type of wood was graded according to the quality of fire it could pro-duce. Fire was essential to our people for cooking and heating. Generally fires were kindled by drilling a cedar shaft into a notch cut in the edge of a dry cedar or cottonwood slab. Sparks would fall into a pile of dry tinder below the notch. When travelling, a smouldering cedar-bark rope carried in a clamshell provided a convenient "slow match". And the spongy material inside a rotting log was used to keep a fire slow burning in order to provide a light when required. (Elliott 1980.)

Medicinal Plants

In the past most common ailments could be cured by some herbal remedy. It was mostly women who knew the plants and how to prepare them as medicine. In some cases knowledge of certain medicines was known only by a person with special power. Much of this knowledge is now [as of 1980] lost to us all. But we still know of the many common remedies passed from generation to generation within a family.

Our ancestors learned of the properties of many medicinal plants by watching injured animals to see which plants they ate or rubbed against their wounds.

Many of the medicinal plants used by our Saanich ancestors are known to contain useful drugs and vitamins. They were usually used in one of three ways – made into a tea and swallowed, made into a lotion or mashed up and put on the skin, or just chewed and swallowed, or chewed and spat out. (Elliott 1980.)

Elsie Claxton with plants used for her ten-barks medicine.

Elsie Claxton and Violet Williams both stated that all tree barks to be used for medicine should be harvested in the early morning, just at sunrise, from the sunrise side (south or southeast side) of the tree. The old people used to say, "Don't take a big chunk [of bark]." Just take a thin strip (about 3-5 cm x 30 cm), and then the tree will heal up, and the person who is being treated will heal up too, just like the tree, Violet explained.

Medicines were often prepared as mixtures, and each family has its own special recipes. These are considered private property, but are always made up for people who request them. Elsie Claxton described one of her special medicines, which we called "ten-barks" medicine, because it has about ten different kinds of bark as ingredients. It was used to treat tuberculosis, spitting of blood and venereal disease. The medicine came originally from

Ten-barks medicine plant parts ready for boiling.

Elsie's father-in-law. He passed it on to Elsie's husband, and her husband passed it on to her. The medicine contains the following ingredients and is described more completely in Turner and Hebda (1990):

- Long strips of bark from Trembling Aspen (*Populus tremuloides*), Cascara (*Rhamnus purshiana*), Arbutus (*Arbutus menziesii*), Pacific Crabapple (*Malus fusca*), Grand Fir (*Abies grandis*), Pacific Willow (*Salix lasiandra*), Saskatoon Berry (*Amelanchier alnifolia*; can substitute twigs cut in pieces), Bitter Cherry (*Prunus emarginata*) and Flowering Dogwood (*Cornus nuttallii*).
- Twigs of June Plum (*Oemleria cerasiformis*), 30 cm long and cut in pieces.
- Rhizome of Licorice Fern (*Polypodium glycyrrhiza*), about 20 cm long, as a sweetener; Trailing Blackberry leaves (*Rubus ursinus*) can also be used for this purpose.
- Prepare the medicine by boiling all the barks and twigs together until the liquid is dark and very strong. The sick person drinks it over a period of time. Ten-barks medicine has been administered successfully to Elsie's family members suffering from tuberculosis. Elsie described how, many years ago, a young woman with tuberculosis so severe she couldn't walk took this medicine and recovered.

Ritual and Ceremonial Uses of Plants

Many plants were used in our rituals and ceremonies related to our spiritual ceremonies. Our ancestors were aware of the power contained in many plants. Their belief in this supernatural power in plants (and animals) influenced the way they used the resources around them. Our elders today tell us that when our people were one with the universe, we were careful, not wasteful, because we respected everything we lived with. We used the power of many plants to help us keep this harmony. (Elliott 1980).

Left to right:
Elsie Claxton,
Violet Williams and
Nancy Turner.

Recreational Uses of Plants

We usually think of children's toys and sports as recreation activities. A few plants were used to make toys for children: elderberry stems were hollowed out to make blowguns; kelp stems were also used to make blowguns. Of course, young boys had their own-sized bows and arrows.

Wooden dishes and gambling sticks for adult games were made out of yew wood. Maple knots were used to make hard balls for a type of hand-ball game. (Elliott 1980).

Plants Used
by the W̱SÁNEĆ People

In this section, we describe the main plants used by the W̱SÁNEĆ people. They are listed in alphabetical order of their scientific names within the major plant groups. The SENĆOŦEN (Saanich language) names for the plants are shown in two writing systems: the name in capital letters is based on a system developed by Dave Elliott that is used by the W̱SÁNEĆ people; the name in lower case in parentheses is a modification of the international phonetic alphabet. Appendix 1 shows the equivalent symbols that represent sounds in SENĆOŦEN.

Seaweeds (Algae) and Other Marine Plants

Sea Wrack   *Fucus gardneri*
DE,LOŦĆ (*t'əlát'th-əlhch*)
Other names: Rockweed.

This short, fleshy seaweed is an olive-green member of the general class of brown algae. Sea Wrack grows up to 50 cm high. In the water it stands erect, but at low tide, it flops down onto the rocks. Plants branch dichotomously (regularly in two equal parts) and the stems are flat, with conspicuous midrib. Sea Wrack clings to rocks with a small but strong holdfast. The branches end in twinned broad, rounded structures called receptacles. At maturity the receptacles swell up to 6 cm long. They contain a clear gelatinous material.

Sea Wrack is found in the middle and lower intertidal zones, attached to rocks. The species ranges along the coast from Alaska to California and is common along the shores of W̱SÁNEĆ territory.

Traditional Use: The gelatinous substance in the swollen receptacles of DE,LOȽĆ makes a soothing salve when applied to burns and sores. Elsie Claxton recalled that people rubbed this seaweed on their arms and legs to strengthen them. Children like to play with DE,LOȽĆ, stepping on the receptacles to make them pop.

Bull Kelp   *Nereocystis luetkeana*
₭O,EṈ (*qw'á 7əng*)

This brown alga is one of the largest and most commonly seen of all the seaweeds. Its long, slender stalk grows up to 30 metres long and is attached to rocks in the subtidal zone by a stout, root-like structure called a holdfast. The cartilage-like flexible stalk is solid at the base, and becomes hollow and increases in diameter toward the upper end. At the top is a hollow bulb, and attached to its upper surface are two clusters of elongated, flat leaf-like blades. The entire plant is greenish brown.

Bull kelp grows in dense masses, forming underwater "jungles" in the subtidal zone of sheltered inlets and bays throughout W̱SÁNEĆ territory and all along BC's coast. The long stalks with their hollow floats are often seen washed up in tangled piles along the beaches.

Traditional Use: Elsie Claxton noted that this seaweed was common on the beach at Tsawout. The long, slender stems were used to make fishing lines. Dave Elliott noted that the fresh stems were alternately soaked in fresh water and dried by stretching in the sun or over smoke until the colour became very pale. The stems were twisted into ropes. They were strong and flexible when wet, but became brittle when dry. The W̱SÁNEĆ used the broad, flat kelp blades to steam camas bulbs, clams, and the meat of deer, seal and porpoise. They placed the blades under and over the food to provide moisture for steam and flavour for the food.

For medicine, people used to harvest the old "roots" (holdfasts). The best places to harvest holdfasts were far out around the islands, in salt water. The holdfast was scraped like a carrot, then hot water poured on the scrapings. The resulting tea was considered effective for treating tuberculosis, haemorrhaging and "any kind of sickness wrong with your insides", according to Elsie Claxton.

| Red Laver | *Porphyra abbottiae* |

Red Laver
and related species
ŁEKES (*lhə'q'əs*)
Other names: Edible Seaweed.

Red Laver grows as a single, membranous blade secured to the substrate by a small disc-shaped holdfast. At maturity, the blade, which varies in colour from greenish to reddish-purple, becomes broad and irregularly lobed with deeply rippled margins. Fully grown it is 20-150 cm long. When exposed at low tide the plant dries out and turns blackish.

Red Laver ranges along the Pacific coast from Alaska to Mexico and is common on intertidal rocks around the Gulf Islands in W̱SÁNEĆ territory. It grows on rocks in the intertidal and subtidal zones. There are a number of other closely related species.

Traditional Use: Elsie Claxton recalls that as a girl of about 14 years old, she used to go with her family to Pender Island and other islands. There they gathered ŁEKES to sell to the Chinese people in Victoria. They collected the seaweed all along the coastline and on the islands. She remembers going on a trip on Easter Sunday with her mother, father and brother in her dad's boat. They also hunted and fished. They collected large quantities of the seaweed, then brought it home and spread it out on the rocks at the East Saanich Reserve to dry in the sun. The dried seaweed was in big sheets and very brittle. The Chinese people came to the Reserve and to the camps in boats to buy it. Dave Elliott noted that people used to harvest seaweed during the spring low tides, when fronds measure about a metre long. They collected the fronds by hand or by cutting them away from the rocks with a sharp knife. They boiled ŁEKES or ate it fresh with clams. Elsie did not recall that they ate the seaweed themselves. Dave Elliott referred to a similar seaweed that people ate; this bright green seaweed, Sea Lettuce (*Ulva lactuca*), was also called ŁEKES.

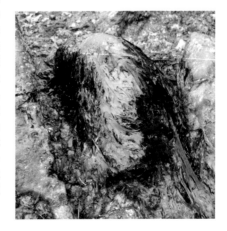

Common Eel-grass   *Zostera marina*
ĆELEM (*chə́ləm*)

Common Eel-grass is an ocean-living flowering plant, a perennial growing from an elongated whitish rhizome and producing clusters of bright-green ribbon-like leaves up to a metre long, sometimes longer. Inconspicuous flowers at the bases of the leaves are pollinated underwater with the help of ocean currents.

Eel-grass thrives in marine bays rooted in mud or sand in the intertidal zone. It occurs in suitable sites all along the coast of British Columbia. Masses of torn-up Common Eel-grass form piles on beaches after a storm.

Traditional Use: The whitish leaf bases of ĆELEM are edible. They are sweet and somewhat salty. The W̱SÁNEĆ people probably ate them, as did other coastal peoples. According to Dave Elliott, the rhizomes and leaves were used to flavour meats steamed in pits. Elsie Claxton said that ĆELEM provided an important habitat for crabs and other marine life.

Fungi, Mosses and Lichens

Tree Fungi
Turkey Tails	*Coriolus versicolor*
Artist's Fungus	*Ganoderma applanatum*
Pine Conk	*Fomitopsis pinicola*
Pore fungi	*Polyporus* spp.

TU,TU,ELEḴEP (*təw'təw ə'ləqəp*; meaning "echo" or "telephone")
Other names for tree fungi: bracket fungi, shelf fungi.

Growing in the form of shelves or brackets on the sides of trees, stumps and logs, the different species have a variety of textures and colours. They do not have the cap-and-stem appearance of mushrooms. Most of these fungi lack stalks, and they are tough and woody. The brackets often have minute holes or pores all over the undersurface, which is often whitish. Within these pores the spores are produced and fall out when ripe. Some of these fungi harm the live trees they grow on by causing the wood to rot.

Artist's Fungus.

Tree fungi commonly grow on decaying logs and on the thick bark of living old-growth trees. They are more commonly found in moist shaded forests, avoiding direct sunlight.

Traditional Use: The W̱SÁNEĆ people apply one general name to the different types of tree fungi: TU,TU,ELEḴEP, meaning "echo", because the fungus is said to reflect a person's voice back when he or she hollers in the woods. Violet Williams recalled that her husband once harvested a huge one from a tree at Goldstream. Many people from the Northwest Coast area consider fungi to bring strong protection against evil thoughts. For example, if the bracket fungus is placed over a door, the fungus is said to protect those of the household from people "talking bad" about them. The fungus is said to answer back for you.

Other reports of the use of fungi include *Tremellodon* (white jelly fungus), which was eaten raw by the W̱SÁNEĆ. A certain species of hemlock fungus, probably Indian Paint Fungus (*Echinodontium tinctorium*), was mixed with cedar bark and alder bark, burned, and powdered to make a reddish pigment for tattoos. A type of willow fungus, possibly *Coriolus versicolor*, was formerly boiled in water to treat babies with convulsions.

Mosses

Step Moss	*Hylocomium splendens*
Electrified Cat's-tail Moss	*Rhytidiadelphus triquetrus*
sphagnum (peat) mosses	*Sphagnum* **spp.**
and others	

KEJI (*q'ə´ch'i7*)

Mosses are small plants that rarely grow more than a few centimetres tall. They are generally bright green, although some are yellowish-green, brownish or reddish. Small spore-bearing capsules are produced from the leaf-like stems. These capsules often grow on slender stalks in the rainy season. The spores are dispersed by air or water.

Mosses readily grow in moist areas throughout W̱SÁNEĆ territory, forming carpets over exposed rocks, decaying logs, on the branches and trunks of trees, and on the ground. They require moisture to grow; in the dry season, many of them become dry and brittle and go dormant. Sphagnum grows in acid bogs and is soft and sponge-like. Many other mosses are stiff and erect, and others hang from trees. Some mosses even live submerged in fresh water.

Traditional Use: The W̱SÁNEĆ call all kinds of mosses KEJI. They sometimes used KEJI to cover food in steaming pits, according to Dave Elliott and Diamond Jenness (1945). They also used them for other household purposes, such as bedding, wiping fish and covering floors. Sphagnum moss was gathered from bogs, such as Rithet's Bog at Royal Oak and formerly in Maber's Swamp – its soft and absorbent qualities made it ideal for use as baby diapers.

Some lichens that are light green are also called KEJI because they resemble mosses in general appearance.

Electrified Cat's-tail Moss, a common carpet-forming moss on the Saanich Peninsula.

Peat-moss (sphagnum) hummock.

Mushrooms
SĆEMSTELIḰ (*sk*ʷ*əm'stəlíq*ʷ; meaning "hat")

There are many kinds of mushrooms to be found in W̱SÁNEĆ territory. Mushrooms come in all different colours, shapes and sizes. Most of them are distinguished by their typical fleshy toadstool appearance, with a flat or rounded cap attached to a stalk. The visible part of the mushroom is the spore-bearing structure, which produces many tiny spores, each with the capacity to grow into a new organism under the right conditions.

Most of the time, mushrooms exist as an invisible network of microscopic thread-like strands under the soil and in decaying plant matter.

Agaricus species.

Mushrooms are most commonly visible on south Vancouver Island in the fall, but some grow in the spring and a few in the summer. The beginning of the rainy season in the fall usually signals the arrival of many different mushrooms. They can be found everywhere, from moist woods to open fields, on the ground or on decaying matter.

Traditional Use: The W̱SÁNEĆ people did not eat SĆEMSTELIḰ in the old days. More recently people have begun to gather and eat field mushrooms (*Agaricus campestris*) and other edible kinds.

Warning: Many mushrooms will make you sick if you eat them, and some are deadly poisonous. Never eat mushrooms unless you are sure of their identity and know them to be safe. There are no good rules to distinguish poisonous from edible types. For those who want to find out more about mushrooms there are many excellent books. Mycologist David Arora has written two that are very good: *Mushrooms Demystified* (1986) and *All that the Rain Promises and More* (1991).

Foliose Lichens
Lung Lichen
Waxpaper Lichen
SMEXDÁLES (sməx̱t'aləs)

Lobaria pulmonaria
Parmelia sulcata

Lung and Waxpaper lichens exhibit a leaf-like form. Lung Lichen is bright green on top when moist and brownish-white underneath. It dries to a light grey-green colour. Its texture resembles lung tissue, hence its common and scientific names. Waxpaper Lichen is grey on top and black underneath.

Lung Lichen grows on the limbs of Bigleaf Maples in moist forests along rivers, such as at Goldstream. Waxpaper Lichen is common on south Vancouver Island and is especially abundant on Garry Oak trees around the Saanich Peninsula.

Traditional Use: Both lichens are called SMEXDÁLES and used medicinally. Elsie Claxton noted that the medicinal properties are apparently dependent upon the type of tree on which the lichen grows. Some kind of lichen was formerly used as a type of birth control. Elsie said that if a woman drank about a gallon of tea made from this lichen she would never have another baby. The hair-like, light-green "old man's beard" lichens, *Alectoria sarmentosa*, *Ramalina reticulata* and *Usnea* spp., were simply called KEJI, the same as tree mosses. Dr Earl Claxton Sr said that his mother, Elsie, told him that these lichens could cause you to become confused and disoriented if they happen to brush across your face when you are out in the woods.

Lung Lichen.

Ferns and Fern-Allies

Lady Fern
LEḴLEḴÁ (ləqləqéy')

Athyrium filix-femina

This large, delicate looking fern grows in dense clumps to a metre or more tall. The fronds spread upward from a thick black rootstock. They are finely divided two or three times into 20 to 40 pairs of leaflets. The general shape of the fronds is lance-like: broad in the middle and tapering at both ends. The spore-bearing structures, called sori, are attached to the underside of the leaflets and look like small, dark spots.

Lady Fern grows only in wet places. It is common throughout W̱SÁNEĆ territory and thrives on the rich soil of the wooded river flats at Goldstream River. It can be found in wet forests, swamps, thickets, landslide tracks, stream banks, gullies, meadows and clearings. Lady Fern often grows with Skunk-cabbage under Western Redcedar and Red Alder.

Traditional Use: LEḴLEḴÁ fiddleheads were said to be good for treating tuberculosis. A tea was made from the shoots and drank over a period of time. Elsie Claxton also stated that the fiddleheads were a good medicine for many things.

Both Violet Williams and Elsie Claxton used a similar name, LEQLEQÁ, for Spiny Wood Fern (*Dryopteris expansa*), which grows on rotten logs in rich woods at Sooke and Jordan River on the southwest shore of Vancouver Island.

Giant Horsetail   *Equisetum telmateia*
SXEMXEM (s̲x̲ə′m′x̲əm′; XEM means "heavy")

Giant Horsetail has separate fertile (spore-bearing) and sterile (vegetative) stems, both of which are hollow and jointed, with a skirt of small fused pointed leaves at each node or joint. The fertile stems are whitish, unbranched and somewhat fleshy. They bear an elongated cone-like structure at the top of the stem that produces the spores. The hollow bright-green sterile stems branch out like the spokes of a wheel at each node. The stem surface is ridged. Plants grow up to two metres in height, and feel scratchy to the touch because of the presence of silica in the cell walls. A smaller species, Common Horsetail (*Equisetum arvense*), is sometimes called SXEMXEM, too.

Giant Horsetail occurs in wet places such as stream banks, ditches, seepage areas and gullies. It often grows near still or running water where it tends to form dense patches. It can be found throughout W̱SÁNEĆ territory and all along the coast of northwestern North America. It is particularly common at Goldstream.

Traditional Use: Elsie Claxton and Violet Williams both recalled that the spore-bearing stems of SXEMXEM were often eaten in the spring, and were quite tasty. The scaly leaves were peeled away from each node, and the inner part was eaten raw. They were also sometimes boiled before being eaten, according to Christopher Paul.

The scratchy stems were occasionally used as sandpaper to polish wooden objects, such as feast dishes and knitting needles, and to smooth the wood of canoes. The roots were sometimes woven into black designs on baskets.

The young shoots of SXEMXEM were believed to be good for the blood (Turner and Bell 1971).

Scouring-rush
X̱,ḰÁL (*x̱wqw'əl*)
Other name: Rough Horsetail.

Equisetum hyemale

Stems of Scouring-rush are unbranched, hollow and green. They have little ridges along their length and grow to a height of 1.5 metres. At the base of each node, there is a small blackish whorl or ring of fused leaflets. The stems arise from dark rhizomes (underground horizontal stem-roots). A soft pine-cone-like structure, 2.5 cm long, grows at the end of a mature stalk and produces spores by which the plant reproduces.

Scouring-rush is common in moist to wet places such as stream banks and riversides, open sandbars, and in ditches, fields and flooded forests. On the Saanich Peninsula it often inhabits the shade of damp thickets of roses and Red-osier Dogwood.

Traditional Use: The rough, silica-impregnated stems were used like sandpaper to polish wooden objects such as canoes, utensils and knitting needles. The stems were also used to scrub out pots, like steel-wool scouring pads.

X̱,ḰÁL was an important medicine for the W̱SÁNEĆ and other First Peoples. For example, Violet Williams said it was used as a medicine for a sore throat. Singers used to drink a tea from X̱,ḰÁL to give them a good, clear voice, especially for singing at Indian dances that could go on for many hours.

Elsie Claxton noted that this species is very similar to SXEMXEM (Giant Horsetail).

Warning: As with many teas made from wild plants, Scouring-rush tea can be harmful if too much is consumed. The stalks are not edible.

Licorice Fern
ṮESIP (*tl'əsíp*)

Polypodium glycyrrhiza

This small fern grows in patches on mossy tree trunks or rock faces. Licorice Fern fronds remain green all winter long, but dry out and turn brown in the summer. They are usually only about 15-25 cm long. Individual fronds with yellowish stalks grow from long, branching greenish-yellow rhizomes. The frond is divided into pointed leaflets, each 2-3 cm long with finely toothed edges. The orange sori are found on the undersides of the fronds, in rows on either side of the leaflet veins.

Licorice Ferns can be found throughout W̱SÁNEĆ territory, and at low elevations throughout the Northwest Coast. They occasionally grow on wet, mossy ground, but usually on rock faces or tree trunks and branches, especially those of Bigleaf Maple and Red Alder.

Traditional Use: The fleshy rhizomes of ṮESIP have a strong, sweet licorice taste and were eaten fresh or dried in the sun for winter use, according to Christopher Paul. They were used as a sweetener and appetizer.

Elsie Claxton also mentioned the sweet rhizomes that were used to sweeten tea and improve the taste of bitter medicines. Violet Williams noted that the ferns grow on trees and rocks, and could be gathered from anywhere. Elsie believed that the rhizomes were more effective when harvested from tree trunks, especially that of Bigleaf Maple. Violet mentioned that ṮESIP was particularly prevalent on the maples at Goldstream.

ṮESIP was also said to be good for treating colds, coughs, sore throats and respiratory ailments like tuberculosis. Christopher Paul and Dave Elliott noted that they were also used for stomach trouble. For coughs, a piece of the rhizome is chewed and the juice swallowed.

Licorice Fern fronds and rhizomes.

Sword Fern
SȾXÁLEM (*sthx̱élə̓m*)

Polystichum munitum

Sword Fern has large, dark-green fronds that grow in clumps from a large root-stock. The plants are evergreen, since the new fronds emerge and mature before the previous year's fronds start to die back. The fronds, which can reach a metre or more in height, stand stiffly up from the crown. The stipe is greenish and densely scaled. The blade has numerous narrow pointed and toothed leaflets arranged in a feather-like fashion. The young fronds begin curled, then gradually unfurl and expand. Small, brownish sori are found in rows on the undersides of some fronds.

Sword Fern is a coastal species that thrives in the shaded humus of damp co-niferous forests. It often grows with Western Redcedar and Red Alder on nutrient-rich seepage sites. In such sites, fern clumps may completely cover the forest floor. They are found throughout W̱SÁNEĆ territory.

Traditional Use: In springtime, SȾXÁLEM rhizomes were dug up, cleaned and cooked in open fires or pits. The cooked rhizomes were then peeled and eaten, usually with seal or bear grease or salmon eggs. The fronds were used to line pits for cooking root foods and to line boxes and baskets that were to hold food.

Elsie Claxton recalled that clams were covered by Salal branches and SȾXÁLEM fronds before being pit-cooked and before they were roasted on skewers over the fire.

Dave Elliott noted, "Before the first dance of a new dancer, Sword Ferns were scattered on the floor by the two or four men who had picked them." (See also Turner and Bell 1971.) Because of their use in these initiation rites and in bathing rituals, SȾXÁLEM are considered to be sacred and should be treated with great reverence. Some say not to use them in cooking.

Bracken Fern
SEḴÁN (*səqéen*)

Pteridium aquilinum

Bracken Fern often forms dense patches from deep, branching black-skinned rhizomes. The stiff, straight fronds can stretch up to two metres tall. The upper part of the frond is roughly triangular in shape, with twice divided segments. The

edges of the frond segments roll under and sometimes have small spore-bearing sori tucked under them. In spring, the fronds thrust out of the ground, looking like curved shepherd's hooks. They have an almond odour, because of the presence of cyanide. The fronds turn light brown and die back in the wintertime.

Bracken Fern is found in a wide variety of habitats and grows especially densely on disturbed sites such as roadsides and soils scorched by fire. The rhizomes grow deeply enough in the soil to avoid most fire damage. This fern also grows in wet forests and places with strongly acid soils, such as bogs and some lakeshores.

Traditional Use: Violet Williams recalled that people once used the large fronds of SEḴÁN as a surface on which to clean fish. The fronds are coarse enough that the fish doesn't slide around while it is being cleaned. Then, when people went to smoke the fish, they used the SEḴÁN fronds to burn as the first fuel on the smoking fire. This was said to give the fish a nice red colour.

Violet Williams also recalled that long ago, people ate large quantities of SEḴÁN rhizomes. The rhizomes were cooked in a pit oven or roasted over a fire until the outer skin and the tough central fibres could be removed. The remaining part, whitish and starchy, was eaten with fish eggs or some kind of oil because it

Bracken Fern rhizomes.

was said to cause constipation on its own (Barnett 1955). Violet Williams never ate the rhizomes herself, and people do not eat them at all today.

The Nanaimo people were said to collect the central fibres of bracken rhizomes, dry them, and used them as a fuel for a slow match, or for torches.

The rhizomes were dug in the late fall or winter, and were always eaten fresh because they do not keep well. Both Violet Williams and Elsie Claxton remembered a story about a man in Nanaimo who was warned not to eat the SEḴÁN rhizomes in the summertime. He ignored the warning and kept eating them and eating them. In this way he got snakes inside him. You should only eat SEḴÁN rhizomes in spring or fall, never in summer, or you'll become "snaked." One version of this story was recorded by Diamond Jenness (n.d.) from a W̱SÁNEĆ person, sometime around 1930.

Snake Island

The Indians always warn their children not to eat too much fern root [bracken] without oil, otherwise they will get something in their stomachs.

Once a boy took no notice of their prohibition, but constantly carried fern root into the woods, roasted and ate it. Soon his stomach became very hard (constipated). As he lay on his back in the woods, snakes entered him. He was ashamed to tell his people, but they discovered what was the matter with him and abandoned him on what is now Snake Island. There the snakes would crawl out of his body when he was lying quiet, but rush back inside him the moment he moved.

His kinsman went over to the island and called to him from the canoe, "How far do the snakes go away from you?"

The boy answered, "Just a few yards."

"Do you think you could run and jump into the canoe before they caught you? For they are afraid of water."

"I'll try."

When the snakes were at their farthest he ran and leaped into the canoe. The snakes dashed after him, but fell into the water and returned to the island. Thus the boy recovered. But ever since then, Snake Island [near Nanaimo] has been overrun by snakes.

Warning: Eating Bracken Fern fiddleheads is not recommended, because in some parts of the world they are linked to stomach cancer.

Cone-Bearing Trees

Grand Fir *Abies grandis*
DEW̱I,EȽĆ (*t'əxwi-ílhch*) or SḰEMÁYEK̲S (*sqwəméy'əqs*)

Grand Firs are straight trees that grow up to 80 metres tall. The bark is grey-brown, and in young trees it is smooth and often dotted with resin blisters that contain strong-smelling liquid pitch. As the tree ages the bark becomes rougher and more ridged. The flat needles end in blunt points with a notch. They are dark green on top, and have two white lines on the underside running their length. The branches are characteristically flat, with the needles spreading horizontally along the twigs. Pollen cones are small and yellowish. The cylindrical seed cones, borne near the top of the tree, stand upright up to 10 cm long. They release winged seeds in late summer, when the cones break up and the scales drop to the ground, leaving only a slender central core on the branch.

Grand Fir grows in moist coniferous forests, on the drier side of Vancouver Island and on the Gulf Islands, throughout W̱SÁNEĆ territory. It is often found with Douglas-fir, but is usually not as common. Grand Fir also occurs on river flats and dry slopes. Elsie Claxton reported that there used to be many on Pender and Saturna islands. Amabilis Fir (*Abies amabilis*), a close relative of Grand Fir, inhabits mountainous regions and the wet west coast from Jordan River northward. Amabilis Fir has short needles that grow in dense rows along the tops of the twigs.

Traditional Use: DEW̱I,EȽĆ boughs were considered quite effective for harvesting herring spawn. Many elders have recalled that the herring used to spawn abundantly in Saanich Inlet, but they haven't been coming there for many years now.

The blistered bark of Grand Fir.

Dave Elliott noted that the liquid pitch from this tree was rubbed on canoe paddles and other wooden articles as a protective coating. It was then scorched to provide a good finish (see also Barnett 1955). Strait-grained "balsam" (DEW̱I,EȽĆ) wood was sometimes bent into a horseshoe shape by steaming it, then lashed into that shape until it dried, fitted with a bone barb, and used as a halibut hook (Jenness 1945).

A tea was made from DEW̱I,EȽĆ bark to treat colds, stomach problems, ulcers and tuberculosis. The liquid pitch from the bark blisters was used medicinally in a number of ways. To soothe a fever or purify the blood, a person could dissolve a small amount of pitch in hot water sweetened with a little sugar, and drink this infusion over a period of about a month.

The bark of SḰEMÁYEK̲S (Grand Fir or possibly Amabilis Fir) was an ingredient in Elsie Claxton's special medicine (see pages 30-31). A strip of the bark was harvested in the morning from the sunrise side of the tree and cut into pieces. It was collected up on the ridge above Tsawout.

Christopher Paul and Dave Elliott noted that an infusion of DEW̱I,EȽĆ root bark made an excellent hair tonic for falling hair and dandruff. It was prepared by pounding the bark from the roots, then steeping it in warm water. The resulting fluid was then rubbed into the scalp.

Chris said that the pitch of DEW̱I,EȽĆ and other conifer species, when mixed with venison suet, made an ointment that could cure psoriasis and other skin diseases, or a salve for cuts and bruises (Turner and Bell 1971).

Yellow-cedar
POŚELEḰ (*páshələqw*)
Chamaecyparis nootkatensis

This cypress-like tree grows up to 50 metres tall. The trunk often twists slightly, and older trees have a ragged appearance. In younger trees the top droops, like that of Western Hemlock. The flat branches hang vertically, appearing limp. The bark is greyish, with long shallow furrows, similar to that of Western Redcedar. The tiny scale-like leaves are blue-green. The yellowish pollen cones are only about 4 mm long. The seed cones are spherical, about the size of a pea or small

Yellow-cedar: A small tree (above), a branch laden with pollen cones (below left) and trunks with peeling bark.

marble, and borne singly, scattered over the branches. When young they are light green and look to be covered in a dusting of fine powder, but as they mature they turn brownish-grey with flat, woody scales, and they release small winged seeds.

Yellow-cedar grows in damp, often rocky areas usually at upper elevations. These include avalanche chutes and ridge tops up to the timberline at mid to high elevations. It occasionally grows at sea level on the west coast of Vancouver Island. It does not generally occur in W̱SÁNEĆ territory, but people would have encountered it in the higher mountains of the Malahat and Jordan River areas.

Traditional Use: POŚELEḴ wood has many uses. The tiny paddles Indian dancers wear on their dancing shirts are made from it. The paddles are about 4 cm long, and are attached by the dozens on the front and back to create a rhythmic rattling sound during dancing. The wood is also used for making real full-sized paddles (see Barnett 1955). POŚELEḴ wood is light, and is also good for making knitting needles and spoons. It was also used formerly for making spindles and spindle whorls for spinning wool. Elsie Claxton and Violet Williams used to use these implements to spin their own wool. These spindles were portable, and both women recall spinning wool while out on the water in canoes. Feast dishes were also sometimes made from large POŚELEḴ logs. In the past the Saanich people obtained POŚELEḴ hunting bows from the mainland through trade; these were strengthened with twisted deer sinew set in a groove and then covered with sturgeon glue (Jenness 1945).

 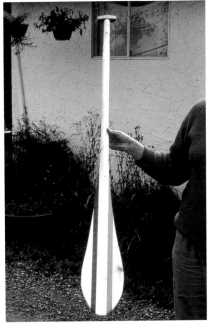

Tiny paddles attached to a dancer's' shirt (left) and a full-size paddle, all made of Yellow-cedar wood.

The inner bark was used like SLEWI (Western Redcedar inner bark) for weaving baskets and mats, and as a thread or string, like Raphia. Since POŚELEḰ is found mainly at upper elevations, people in the old days would have obtained it during hunting-and-gathering expeditions into the mountains.

Seaside Juniper   *Juniperus maritima*
PEṮEṈIŁĆ or PEPEṮEṈ (*p'ət'thəngi-ilhch* or *pəpət'thíng*; meaning "skunk" or "rank-smelling tree")

Seaside Juniper is a small tree that grows up to 15 metres tall and is often gnarled. It has shaggy bark and drooping, greenish blue branches. The leaves are tiny and scale-like. The male and female cones are produced on separate trees. The female cones are berry-like, about the size a pea, and bluish green to black with a waxy coating. The foliage and berry-like cones have a pungent, spicy smell, especially when crushed. Seaside Juniper lives along the dry south coast of British Columbia and southward into the United States. It grows on rocky headlands on the Saanich Peninsula, and as far north on Vancouver Island and the Gulf Islands as Yellow Point. Violet Williams thought that the old people might have planted one tree that grows at Patricia Bay.

Traditional Use: This juniper is an important spiritual and protective tree. When there is sickness or death in the family, the fragrant, strong-smelling branches of PEṮEṈIŁĆ are hung around the house or placed under the mattress to protect people against illness and "to drive the germs away" according to Christopher Paul. Elsie Claxton said about this tree, "It takes all sickness away."

Note: Until a few years ago this tree was classified as Rocky Mountain Juniper (*Juniperus scopulorum*), but recent botanical studies of the plants on the Saanich Peninsula helped show that the juniper here on the coast is a different species than the one that grows in the interior mountain ranges (Adams 2007).

Juniperus maritima (Seaside Juniper) growing on the shore at Tsartlip. This tree and those growing with it have been designated as the official living specimens representing this newly described species.

Sitka Spruce
TŦḴÁ,IŁĆ *(t'thq'é7-ilhch)*

Picea sitchensis

Sitka Spruce is a straight-growing tree with bluish-green boughs and greyish, scaly bark. The branches are stiff and outspreading in young trees, and somewhat drooping in older trees. Young Sitka spruce trees closely resemble young Douglas-firs. But like all spruces, Sitka Spruce has sharply pointed needles that can be rolled in the fingers because they are four-sided rather than flat – just touching the branches will enable anyone to distinguish this as a spruce. The male cones are small and reddish, and the female cones are light brown, around 10 cm long, and have thin, papery scales that are wavy at the tips.

Sitka Spruce occurs sporadically on the Saanich Peninsula, in shady, forested, moist ground. It is much more common in forests along the west coast of Vancouver Island from Sooke to Jordan River and beyond.

Traditional Use: This spruce is very pitchy, and the W̱SÁNEĆ people may have used the pitch, as they did pine pitch and fir pitch, as a caulking and waterproofing agent and as a salve for slivers, wounds and skin infections. Diamond Jenness (1945) said that people often made halibut-fishing line from spruce roots, with hooks of hemlock knots and bait of octopus. Elsie Claxton and Violet Williams knew of no specific use for it.

Note: Tim Montler points out that TŦḴÁ,IŁĆ is almost the same as ṮḴI,EŁĆ, the word for Western Hemlock, and so might be incorrect.

Lodgepole Pine   *Pinus contorta*
White Pine   *Pinus monticola*
ḴÁYÁLEŚIȽĆ (*qw'əyé7ləs-əlhch*; meaning "dancing-plant/tree")

Lodgepole (Shore) Pine is a relatively small tree that can grow straight, but often has a crooked, bushy form. It has dark, furrowed bark, and its long needles grow in pairs. Its small pollen cones produce clouds of pollen in the spring, and its short, clustered seed cones usually remain tightly closed until heated by fire.

White Pine is a tall, straight, attractive tree that grows up to 30 metres tall, sometime taller in good sites. Its bark is whitish and smooth and its needles, up to 10 cm long and bluish green in colour, are borne in clusters of five. The pollen cones are yellowish and clustered, usually under one centimetre long. The seed cones are borne on the upper branches. They are 15-25 cm long and cylindrical, with scales opening out in warm fall weather to release winged seeds. The tips of the scales are often covered with pitch.

Lodgepole Pine grows quite commonly around the Saanich Peninsula and the Gulf Islands, often in places with poor soils, including bogs, rocky hilltops and exposed coastline. White Pine is not very common in W̱SÁNEĆ territory, but you might encounter scattered individuals in moist forests, usually at mid elevations. It was formerly more common, but has been seriously impacted by the introduced White Pine Blister Rust.

Traditional Use: Dave Elliott noted that the juicy inner bark of Lodgepole Pine was eaten fresh, or dried in cakes. It was pulled off the tree in strips in the springtime. It is unclear, however, if this use pertained to the W̱SÁNEĆ people or to peoples of the interior, who commonly ate this food (see Turner and Bell 1971).

Both White and Lodgepole pines are quite pitchy. The latter's pitch was used to fasten arrowheads onto shafts, according to Christopher Paul. Pine sap, like

Young cones of White Pine.

Grand Fir sap, was mixed with deer tallow and applied to the skin for psoriasis and other ailments.

An old W̱SÁNEĆ legend relates how the mythical cannibal woman sealed the eyes of the children she stole with pine or fir pitch so that they wouldn't be able to see where she was taking them.

Notes: K̲ÁYÁLEŚIⱢĆ means "dancing-tree", similar to the name for Trembling Aspen (see page 75). For aspen, the dance is in the trembling leaves. For these pines, it is apparently in the branches – if you place a needled twig upside down on a flat surface, it will move around with the least vibration, resembling the motions of a traditional dancer, according to Dr Arvid Charlie, *Luschiim*, of Duncan.

Douglas-fir   *Pseudotsuga menziesii*
JSÁY (*ch'sey'*)

Douglas-fir trees have been known to reach an age of well over 1000 years. They have spreading needle-covered branches that stretch far out from the trunk. Older trees can reach 90 metres tall and be more than two metres across the base. The bark is grey and smooth in young trees, and as it matures, it thickens and develops large deep grooves. Mottled reddish-brown in colour when old, the thick bark protects the trees against fires. There are scattered living ancient Douglas-firs in W̱SÁNEĆ territory that bear charcoal scars from many fires. The needles, 2-4 cm long, are slightly pointed and feel soft to the touch, not like Sitka spruce. They are arranged in a bottle-brush fashion on the twigs. Grand Fir and the other true firs (*Abies* spp.) have notched needle tips and flatter boughs. A Douglas-fir has two kinds of cones: small reddish brown male cones that produce clouds of yellow pollen when they are ripe, and larger woody seed-bearing cones, 5-10 cm long, with three-pointed little bracts protruding from the cone scales – a distinctive feature of Douglas-fir cones.

Douglas-firs range throughout most of the southern third of British Columbia, and they are very common in W̱SÁNEĆ territory. The tree favours dry sites at medium to low elevations, though it is also known to reach timberline. It is often associated with exposed rocky outcroppings and areas visited by fire.

Traditional Use: JSÁY bark is considered to be a top-quality fuel, because it burns with a hot, smokeless flame. The wood was used to make spoons, seal harpoon shafts, fire tongs and many other implements (Barnett 1955). The dense knots in the wood were moulded into curved halibut hooks by steaming them, placing them in a hollow kelp stem overnight and then bending them to the proper shape. To keep their shape after moulding, the hooks were rubbed with tallow. Homer Barnett (1955) also reported that rotten JSÁY wood was sometimes used in the tanning of animal hides, and that the poles and branches were woven together to make salmon weirs. The strong-smelling, sticky pitch was used as a cement to patch canoes and water containers. It was also used as a salve for wounds.

The following story – part of a longer account of XELS (the Transformer) and his activities – is about how JSÁY originally got its pitch, and why Arbutus sheds its bark every year (Jenness n.d.):

How Douglas-fir Got Pitch

Pitch used to go fishing before the sun rose, and retire to the shade before it became strong. One day he was late and had just reached the beach when he melted. Other people rushed to share him. Fir [Douglas-fir] arrived first and secured most of the pitch, which he poured over his head and body. Balsam [Grand Fir] obtained only a little; and by the time Arbutus arrived there was none left.

Arbutus said, "I shall have to peel my skin every year and have a good wash to keep me clean."

But just then XELS appeared and said, "You shall all be trees and Fir shall be your boss."

So now the Arbutus sheds its bark every year, and [Douglas-]fir has more pitch than any other tree.

Pacific Yew
T̲E̲N̲KÁŁĆ (*tl'əng'q'-ílhch*)
Other name: Western Yew.

Taxus brevifolia

Pacific Yew is a small, stocky tree, with a furrowed, gnarled trunk, usually not reaching a height of more than 15 metres. The branches spread outwards, and are often twisted and fluted. The bark is a reddish colour, often papery and sloughing off the trunk in shreds. The needles are flat and pointed, dark green on top and lighter green beneath. They extend horizontally from the twig on either side, making the branches distinctively flat. Male and female reproductive structures are borne on different trees. The male pollen cones are small and inconspicuous. Instead of seed cones, the female trees produce arils, bright red fleshy cups containing a single seed. Birds eat these seeds, but they are poisonous to humans.

Western Yew grows in mature, moist, shaded woods, often with Douglas-fir, Western Redcedar and Western Hemlock. It grows sporadically at low to middle elevations, usually under the canopy of the forest, throughout W̲SÁNEĆ territory.

Traditional Use: Both Elsie Claxton and Violet Williams called this tree "ironwood". T̲E̲N̲KÁŁĆ wood is very tough, and for this reason it was important in W̲SÁNEĆ technology. It was used for weapons and implements that need to be strong and resilient, such as bows, harpoons shafts, halibut hooks, paddles, wedges, and even combs and gambling discs. Christopher Paul said that the fine-grained red heartwood was considered stronger and superior to the white sapwood. Homer Barnett (1955) noted that pegs of T̲E̲N̲KÁŁĆ dipped in boiling pitch were used to fasten box corners together to make a tight seal. Diamond Jenness (1945) reported that young W̲SÁNEĆ girls formerly used yew twigs to remove underarm hair.

Yew bark is quite famous now as a source of a potent anti-cancer drug called Taxol. Tim Montler recalled that when news of this discovery came to light in the early 1980s, both Elsie and Violet confirmed it by saying, "Oh, yes, it's good for cancer." In some places Western Yew has been overharvested, and many elders and carvers say that it is hard to find yew trees any more. The W̲SÁNEĆ and other

Yew-wood mat creaser
in the shape of a bird
(RBCM 1177).

First Peoples have used ṮEṈḰÁȽĆ bark (and wood in some cases) as medicine for a long time – probably centuries. The bark is an ingredient in Violet's special four-barks medicine that she used to make for those who had internal problems like ulcers and liver ailments.

Warning: All parts of the Pacific Yew tree can be toxic. Although it is used medicinally, it should never be administered or taken by people who do not know how to prepare it properly. Always check with a qualified herbal specialist or physician before trying any kind of medicine from plants.

Western Redcedar   *Thuja plicata*
Tree, wood, cedar canoe, cedar post: XPÁY (*x̲péy'*)
Plank: XEXPA̱ (*x̲ax̲péy'*)
Inner bark: SLEWI (*slə'wi7*)
Bark (general), outer bark: JELA̱ (*ch'əléy'*)
Boughs, branches: XPÁY, ÁSES (*x̲pey'esəs*)
Withes (for rope): STESTÁSES (*st'həstésəs*)
Withe rope: XA̱EX̲TEN (*x̲ə'y'əxwtən*)
Roots: ĆEMLEX̲ (*kwə'mləxw*)

Western Redcedars reach to 60 metres in height. The top of the tree droops, like that of Western Hemlock. Mature trees are often buttressed at the base with wide ribs of wood running out from the main trunk. The branches tend to droop slightly before turning upward, in somewhat of a "J" shape. The bark is grey to reddish brown, appearing as fibrous vertical strips. The tiny leaves are scale-like on the branchlets, overlapping like shingles on a roof. The boughs are a glossy light green to brownish green. In late summer and early fall, the oldest branchlets turn yellow or reddish and drop off. The reddish male pollen cones are small and numerous. The seed cones are egg shaped, growing up to one centimetre long, with 8-12 scales per cone. They are usually clustered together on the outer boughs. As the seed cones mature they turn from green to brown and point upward. The seeds are winged when shed.

    Western Redcedar thrives in moist to wet soils at low to mid elevations. It grows best in shaded forests near sites that are often wet from seeping or flooding,

Elsie Claxton with cedar branches.

but is also found in drier areas where the soil is shaded. Widespread in W̱SÁNEĆ territory, this is one of the most important of all the trees.

Traditional Use: According to Elsie Claxton and Violet Williams, XPÁY was the only kind of wood or tree used for canoes. In the past coffins were made of cedar, in the style of a bent-wood box, because cedar wood never rots. These boxes were placed up in trees. XPÁY is also used for house construction (permanent winter homes and temporary summer houses), salmon drying racks and many other purposes. Dave Elliott said, "Cedar wood is extremely rot resistant and easily split. Logs were split into large long planks and used to make our Big Houses. Also, cedar wood was made into dugout canoes, totems, bentwood boxes, cradles, coffins, drum frames, masks, herring rakes and canoe bailers. The young branches would be twisted to make reef-net frames and ropes."

Sheets of JELA̱ (outer and inner bark together) were used for roofing and temporary shelters. Both Elsie and Violet recalled that their mothers used to make baskets out of ȻEMLEX̱. These baskets were used to hold many items such as clams and clothing, and were all different sizes and styles. The most common was about 30 cm across and about 20 cm deep. Violet's mother used to dig up ȻEMLEX̱ on the hill above Westholme where they lived, but good roots were available in many different locations. ȻEMLEX̱ were dug from PENÁW̱EN through ĆENHENEN (May to July), at the same time as the JELA̱ is ready to pull off. ȻEMLEX̱ was harvested by digging a small area, locating a good root, and then pulling it out, as far as it would go without breaking. The roots were different sizes, but mostly about the diameter of a Tule stalk (about 1 cm). ȻEMLEX̱ were pulled through the teeth to skin them.

Immature (green) and mature (brown) Western Redcedar cones.

Bark (wide strips) and roots (narrow strips) of Western Redcedar prepared for weaving.

Homer Barnett (1955), too, reported on the use of cedar roots by Island Salish peoples: Basket makers would seek a straight tree with even branches and few knots, and dig a pit, sometimes as deep as two metres, to collect the finest quality roots. These were used for red designs in woven baskets as well as for actual weaving of baskets, hats and mats.

SLEWI (the inner bark), like that of Bitter Cherry, is used for sewing things, and for making ropes (see page 77). Strips of the bark were used for basket weaving and for threading through cooked clams for hanging and drying. The Quw'utsun' used cedar bark for tanning fish hooks (imparting tannin as a preservative). People pulled long strips of bark off straight trees with few lower limbs. The best time to harvest SLEWI was usually in the early summer, but it depended upon the weather and conditions of a particular year. SLEWI harvesters were careful to take only one or two strips from each tree, so that they wouldn't kill the tree. The outer bark was removed and the inner bark was coiled into bolts and dried for later use. Dave Elliott noted:

> When it was used, first it was soaked in water and then beaten with a whalebone or yew-wood beater over the edge of a board to separate the fibres. These fibres were known as shredded cedar bark and were used to make mats, hats, clothing, baskets and fishing line. The leftover pieces had many household uses – bedding, diapers, towels, tinder, work aprons, threading clams, covering a drummer's hands in the winter dances, and as a slow match.

Unpeeled withes were used for baskets. They were spaced about a centimetre apart and then formed into a square box with a slightly curved bottom, straight sides and a handle. The crab basket – one type of basket made from withes – was about 60 cm across at the base, narrowing to 30 cm at the mouth, and set in the ocean to keep the crabs alive. Cedar withes were harvested in PENÁW̱EN̲ or ĆEN̲T̸EḴI (May or June). Violet Williams remembers her mother splitting withes in half or sometimes using them whole. Elsie Claxton mentioned that her mother

Western Redcedar tree with a strip of bark taken from it.

used to seek out the straightest withes, such as those that grow along the trail to the nature house at Goldstream.

Cedar withes were also used for ropes and harpoon lines (Barnett 1955). The W̱SÁNEĆ used a giant anchor rope of SȾESTÁSES to hold their canoe to the top of Mount Newton during the rains of the Great Flood. It is said that there is still a coiled cedar-branch rope at the top of Mount Newton, but nobody has seen it recently. You have to be especially clean and well trained to be able to see it.

The W̱SÁNEĆ and Quw'utsun' wove cedar roots with gooseberry and wild-rose roots to make reef nets (Turner and Bell 1971). Cedar withe nets were also strung across small islets in the Finlayson Arm channel to capture ducks. The boughs were also used to catch herring spawn.

Christopher Paul and Dave Elliott noted that cedar-bark neckbands and head-bands were dyed with Red Alder bark and worn by dancers, shamans and all others affected by spirits and needing special protection. Cedar boughs are used by initiates and those in training to scrub their skin during bathing rituals. The boughs are then placed under a rock near the edge of the water. Hunters and others desiring purity scrubbed themselves with cedar boughs.

Elsie reported that a person who worked with XPÁY (making canoes or split-ting planks) could not stay up in the woods for very long because the tree is so sacred. They should leave the woods by mid afternoon. Some people in the old days were said to have special powers that they used to "talk to" a XPÁY log. The log would become so light it could almost float out of the woods. This made it easy to carry.

Violet's mother and father used to tell her that the cedar boughs were men or women. The man was called JAḴELETN (*tsayqələtn*) and the woman SAȾḰAMAT (*sathqwamaat*). Tim Montler noted that these are not the words for "man" and "woman" in either Saanich or Halkomelem, and that they must be traditional proper names, possibly connected to a story.

Western Hemlock   *Tsuga heterophylla*
TḴI,EŁĆ (*t'thq'i-ilhch*)

Western Hemlock trees can grow up to 60 metres tall. The narrow crown droops at the top. Its branches sweep downward, having a feathery appearance. The bark is rough and reddish brown, becoming thick and furrowed as the tree ages. The needles are short, flat and blunt tipped; they are irregularly spaced on the sides of the branches, giving the boughs a flat, lacy look. The pollen cones are numerous and small. The seed cones are relatively small as well, usually 1.5-2 cm long. Young cones are greenish or purplish, but turn light brown as they mature.

A related species, Mountain Hemlock (*Tsuga mertensiana*), inhabits upper elevation forests with Yellow-cedar and Amabilis Fir. It has longer cones that are deep purple when young and its boughs are more brushy-looking than those of Western Hemlock.

Western Hemlock lives in moist, often shady sites. It is usually associated with thick forest humus and decaying wood, but also grows in mineral-rich soils. Although it occurs mostly at low to mid elevations, it is not common on the Saanich Peninsula and the Gulf Islands because of the dry climate. The tree predominates on the west coast, from Sooke northward. Mountain Hemlock occurs in the cool damp forests of the mountains of the Malahat and Sooke ranges west of the Saanich Peninsula.

Traditional Use: According to Homer Barnett (1955), Island Salish people, possibly including W̲SÁNEĆ, collected the inner bark of T̲K̲I,EȽĆ in spring and ate it fresh, sometimes with oil.

Christopher Paul said that fresh T̲K̲I,EȽĆ wood is as easy to carve as cedar, but it becomes too hard after it dries. It was sometimes used to make spoons, roasting spits, spear shafts, wedges and other carved tools.

Elsie Claxton and Violet Williams used to make baskets of Raphia fibres (see page 139) and coloured the strands – one of the dyes they used was made from the bark of a needled tree, apparently T̲K̲I,EȽĆ. They boiled the bark and soaked the Raphia in the solution, which gave it a blackish colour.

Young W̲SÁNEĆ girls used to paint their cheeks with a reddish dye made from T̲K̲I,EȽĆ bark, according to Diamond Jenness (1945).

Flowering Trees

Douglas Maple
BEN,Á,YEŁP (*p'ən7ə7y-əlhp*)
Other names: Rocky Mountain Maple, Vine Maple.

Acer glabrum

Douglas Maple grows as a many-stemmed shrub or a small tree up to 10 metres tall. The young, new branches are reddish in colour, whereas the bark on the mature trunk is greyish. The branches and leaves grow in opposite pairs. The leaves are relatively small, growing up to 10 cm or so across, with three to five pointed lobes and toothed margins. Flowers are small and whitish and grow in clusters at the ends of shoots. The winged fruits, usually in pairs, are 2-4 cm long, smaller than those of Bigleaf Maple, and smooth rather than prickly. The leaves often turn a brilliant reddish-orange in the autumn.

Douglas Maple is not a common tree on the coast, but does occur sporadically in many parts of W̱SÁNEĆ territory. It grows in moist areas under the forest canopy, or along the shoreline, such as at Tsawout, Pauquachin and Mount Douglas beaches. It also occurs on mountainsides at mid elevation.

Traditional Use: The W̱SÁNEĆ people use the hard, whitish wood to make the miniature paddle decorations for dancing regalia. Elsie Claxton said that the whitest wood is obtained from trees growing near the water or beach; the wood of inland trees is said to turn reddish. BEN,Á,YEŁP wood was also used for making knitting needles.

The bark was used to make a medicinal antidote for any kind of poisoning, by simply steeping it in boiling water and drinking the tea. It could also be mixed with Devil's-club bark to make a medicine for sugar diabetes.

Note: The name "Vine Maple" is usually applied to another species, *Acer circinatum*, that occurs commonly in the Fraser River valley.

Bigleaf Maple
ȾȻÁ,EȽĆ (*t'ththé7əlhch*)

Acer macrophyllum

Bigleaf Maple is a tall, spreading deciduous tree growing up to 30 metres tall. The root system spreads widely, from a short, stout trunk up to 1.5 metres in diameter, and the limbs are large, heavy and widely spread. Often on older trees the trunk and limbs are festooned with mosses, lichens and licorice fern. The tree will often grow suckers from the base if the main trunk is cut down. The suckers will, in time, grow into a cluster of new young trees. The leaves of this maple are the largest of any native tree in Canada. Their form is typical for maples: three to five pointed lobes with deep indentations between. The young leaves in spring are a soft yellow green in the sun's light. They soon turn darker and some grow to 30 cm or more across. In fall they turn golden yellow, then light brown, and fall to the earth to be crunched underfoot. The tree flowers in early spring, before the leaves have fully expanded. The small, greenish-yellow flowers hang in dense clusters.

Bigleaf Maples in fall colours along the Gorge waterway in Esquimalt.

Bigleaf Maple flowers (left) and a dry fallen leaf caught in a bush.

Male and female flowers are separate, but occur in the same cluster. Each flower cluster draws pollinating insects until each tree hums like a giant machine. The fruits are rather large, consisting of pairs or triplets of seeds, each with a broad wing that sends the seed spinning to the ground when it is released. The spiny hairs on the seeds can irritate the skin.

Bigleaf Maple grows in moist, rich soils, especially along rivers, streams and floodplains throughout W̱SÁNEĆ territory. Curiously, it is also found on moist rocky slopes often rooted in the rubble at the foot of a cliff. The tree also thrives in disturbed settings along roads and fences. It provides important shelter for many species of wildlife.

Traditional Use: Homer Barnett (1955) noted that in spring, people scraped off the cambium with a stick made of Oceanspray and ate it fresh. It was considered a real treat.

Dave Elliott reported that the wood of ȾȾÁ,EȽĆ can be easily carved, but it is hard and does not warp or crack. It was used to make dishes, spoons, combs, cattail-mat creasers, ceremonial rattles, cedar-bark shredders, adze handles and numerous other tools. It was particularly prized for carving beautiful spindle whorls and canoe paddles.

The large leaves were used for various household tasks. They were sometimes placed in steaming pits to flavour meat, such as seal, porpoise or deer. Dave recalled that they were also used for wrapping fish for cooking, lining baskets and placing under berry-drying racks. ȾȾÁ,EȽĆ wood was also considered a good fuel, because it burns with a hot, smokeless flame.

Some people used the bark and leaves of ȾȾÁ,EȽĆ medicinally as well. For a persistent sore throat, Violet Williams said that you can collect fallen maple leaves that have been caught up on bushes before they hit the ground, and boil them with Scouring-rush and Stinging Nettle roots, then drink the resulting tea. She also noted that the leaves were rubbed on the face of young men at puberty to prevent them from growing thick whiskers.

Saanich Bay, or Tsartlip, is called W̱JOȽEȽP ("Place of Maple Leaves"), according to Dave Elliott.

Red Alder   *Alnus rubra*
Tree: SḴOLṈEȽĆ (*sqwálng-əlhch*)
Edible inner bark: SXAMEŦ (*sx̱a´m'əth*)

This deciduous tree grows up to 25 metres tall. The bark is thin, grey and smooth, and is frequently covered in lichens. The wood and inner bark turn a rusty colour when cut. Red Alder leaves are broad and elliptical with evenly toothed edges. Male flowers hang in catkins from the branch ends. These catkins form before the leaves in the springtime. Female flowers are green, and form into stiff woody cones that remain attached to the tree over winter. The winged nutlets (seeds) are dispersed by the wind.

Red Alder grows in moist woods, along stream banks, floodplains and recently cleared areas that receive direct or slightly filtered sunlight. They often occur in pure stands at low elevations. They are common throughout W̱SÁNEĆ territory. There are many alders growing on the flood plains at the Goldstream Reserve. Elsie Claxton said that they were not there when she used to go there as a child.

Traditional Use: People, especially children, used to eat the SXAMEŦ often. They removed a section of bark and scraped off the moist growing tissue on the outside of the wood. SXAMEŦ is clear and very sweet, and in the springtime really thick. People did not usually bring it home, but just ate it at the tree. Even the young trees have good SXAMEŦ. The old people used to say that when the tide was really low in the springtime, the SXAMEŦ was thick and best to eat. Elsie Claxton said that when the tide goes down, the SXAMEŦ gets "dry", and when it comes up it gets "fat." Dave Elliott said that SXAMEŦ was good for the stomach and was used by the old people for a tonic as well as a food.

SḴOLṈEȽĆ outer bark was sometimes used in cooking pits to steam camas bulbs and give them a red colour.

The colourful bark of Red Alder.

SḰOLṈEŁĆ wood is good for carving items such as dishes and spoons. Elsie Claxton and Violet Williams said that alder wood is the preferred fuel for barbecuing or smoking salmon. Christopher Paul noted that the bark was formerly used to make a dye to colour fishnets, making them invisible to the fish at night. It was put in a big pot or tub and boiled. It was also used for colouring wool and other materials. For face paint, the bark juice was sometimes mixed with red ochre, baked in an oven until dry, then ground into powder. Elsie said that if you wanted to make a dark dye, you must use the bark from an older tree, and to make a paler dye use the bark of a small alder. Alder bark, cedar bark and Indian paint fungus were burned to make a dark red pigment for tattooing, according to Diamond Jenness (1945).

The immature green catkins (when about 1 cm long) of SḰOLṈEŁĆ were nibbled upon as a medicine. The bark is widely known as a medicine for internal ailments like tuberculosis, and as a wash for skin infections. It is usually boiled in water before drinking or applying to skin. Both Elsie and Violet knew it well as a special medicine.

They also stated that rubbing powdered rotten SḰOLṈEŁĆ wood under the arms of boys at puberty (when they start to change voice), will prevent them from having body odour. Elsie once collected some small pieces of rotten alder wood to treat her grandson. Violet mentioned that you could also rub rotting SḰOLṈEŁĆ wood on your feet to prevent bad foot odour.

Arbutus
ḰEḰEIŁĆ (*qwəqwəy-ílhch*)
Other name: Pacific Madrone.

Arbutus menziesii

Arbutus is an evergreen broadleaf tree with a short, stout trunk and smooth, heavy branches. It grows up to 30 metres tall. The bark of the trunk can be relatively thick and scaly, but over most of the tree, it peels and sheds every year. The older bark is bright reddish-brown, and when it peels, the new bark underneath is greenish. The thick, oval leaves are about 15 cm long, somewhat pointed, shiny on the top and dull green underneath. Small, white, urn-shaped flowers bloom in late spring in dense clusters. They mature by late summer into pea-sized orange-red berries that are dry and seedy but not poisonous.

Arbutus grows on dry, rocky sites exposed to sunlight and often together with Garry oak. It also grows with Douglas-fir in deeper soils, where it reaches its maximum height. It is especially common along the coast throughout much of W̱SÁNEĆ territory.

Traditional Use: Elsie Claxton's mother used to prepare an infusion of ḰEḰEIŁĆ leaves and bark by pouring hot water over them. A person drank the infusion to treat a cold and, Elsie added, "for the stomach, too, if there's something wrong with your stomach – any time." It could also be taken orally to treat tuberculosis. The leaves were sometimes chewed raw and the juice swallowed to ease symptoms of a bad cold.

The bark was an ingredient in Elsie's special medicine (see pages 30-31). The bark was usually harvested in May or June when it is easiest to peel off. A strip of bark about 60 cm long and 10-15 cm wide was cut up into chunks and boiled with the other ingredients.

Arbutus trees and berries.

Dave Elliott added that W̱SÁNEĆ people rubbed the leaves on areas affected by rheumatism, and that the Quw'utsun' people rubbed them on burns.

Both Dave and Elsie noted that ḰEḰEIⱢĆ bark was also placed in steaming pits to colour camas bulbs reddish. The wood is hard and brittle, which makes it of little use for carving. But the young branches could be used to make wooden spoons and gambling sticks. A dye from the bark was used for tanning paddles and fish hooks.

According to Chris Paul and Phillip Paul, the W̱SÁNEĆ people have a special relationship with this tree, because during the Great Flood, their ancestors anchored their canoes to ḰEḰEIⱢĆ trees on the top of ȽÁWELṈEW̱ (*lhewəlngəxw*). With its strong, penetrating roots, ḰEḰEIⱢĆ gripped the rock on the mountaintop and kept the people from floating away until the floodwaters receded. Philip Paul said that, to this day, W̱SÁNEĆ people do not burn ḰEḰEIⱢĆ, out of respect for its having saved their lives.

The Flood

Once it rained and rained until the water of the sea began to rise. People twisted cedar branches into a very long rope so that they might anchor their canoes to some rock or stump [Philip Paul said it was an arbutus tree] on the highest mountain if all the land were flooded. The waters stopped just below the summit of ȽÁWELṈEW̱ [*lhewəlngəxw* (Mount Newton)], on Saanich Peninsula, and then receded. Many Indians were drowned, so that today, at a creek near Nanaimo, you can sometimes hear the drowned people still talking about the rising waters.

This story is from Diamond Jenness (n.d.). There is another story from Jenness involving Arbutus on page 57; it explains why Douglas-fir has a lot of pitch and Arbutus has none.

Thin, peeling bark and balding trunks.

Pacific Dogwood   *Cornus nuttallii*
ĆETXIȽĆ (*kwətx̱-ílhch*)
Other name: Western Flowering Dogwood.

This well-known deciduous tree is the floral emblem for British Columbia. Pacific Dogwood has pointed, smooth-edged leaves and smooth, greyish bark. The flowers begin to form in clusters of tight, green buttons in the fall and over the winter. As they mature, the clusters become surrounded by large, whitish, petal-like bracts. The fruits developing from the central flower clusters are bright red and fleshy, each with a central seed. The leaves turn reddish to peach-coloured in the fall.

Pacific Dogwood grows sporadically in open woodlands and edges of meadows throughout W̱SÁNEĆ territory. It can be easily recognized by its flowers in the late spring and sometimes again in the fall.

Traditional Use: The ripe fruits of ĆETXIȽĆ are said to be good for treating pimples and acne when mashed and rubbed on the face.

ĆETXIȽĆ bark was an ingredient in Elsie Claxton's special medicine (see pages 30-31). A strip of bark, harvested in the morning from the sunrise side of the tree, was cut into pieces and boiled with the other ingredients (see Turner and Hebda 1990).

Dave Elliott noted that ĆETXIȽĆ wood was used to make bows and arrows, and the bark was boiled to make a tanning agent and preservative for items such as cedar-bark canoe bailers.

British Columbia's floral emblem (left), and dogwood leaves ready to drop off in the fall.

Pacific Crabapple
Tree: ḰÁ,EW̱IȽĆ (*qəxwi7-ilhch*)
Fruit: ḰÁ,EW̱ (*qé7əxw*)
Other name: Oregon crabapple.

Malus fusca

Pacific Crabapple is a tough, bushy deciduous tree that grows up to 12 metres tall. The spur shoots on the tree are sharp. Older trees have bark that is deeply fissured. The trees and the leaves resemble apple trees in many respects. One difference is that the leaves of Pacific Crabapple, though oval and pointed like orchard apple leaves, often have a coarse tooth or pointed lobe along one or both margins. The edges are also finely toothed. The flowers are white or sometimes pinkish, borne in rounded clusters in the spring.

The small, elongated apples have long stems and grow in clusters of 5 to 15; they ripen in late summer and fall, appearing greenish or yellowish.

Pacific Crabapple grows in moist woodlands and alongside streams, swamps, lakeshores and beaches. It is commonly found around river estuaries, where it can form dense thickets. It occurs throughout W̱SÁNEĆ territory, but due to the loss of wetlands, it is not as common as it used to be.

Traditional Use: Wayne Suttles (1951) reports that crabapples were picked in August and put into cattail bags to ripen for winter. Elsie Claxton and Violet Williams recalled that people used to harvest and eat these "little apples" in large quantities in late summer and fall. They generally harvested the whole clusters. Elsie was very fond of them, but Vi found them quite sour. Vi's mother used to gather ḰÁ,EW̱ and put them in large crocks with just water ("no sugar, oil or anything"). Then, the apples were taken out "if you want to have some kind of a dessert after eating," Vi recalled. Elsie said ḰÁ,EW̱ were gathered while they were still a little green, then taken into the house and "the next day it's ripe." She used to "eat and eat and eat" them. Dave Elliott said that ḰÁ,EW̱ could be boiled in water and Eulachon oil and stored covered in their own juice.

Elsie said that if a ḰÁ,EW̱IȽĆ log was big enough, it could be used as a fuel for smoking and drying salmon. But she and Vi said that most of the time alder wood is used for smoking salmon. Dave Elliott noted that ḰÁ,EW̱IȽĆ wood is very hard and tough, and people sometimes used it to make tools of various kinds, such as the handles of adzes and axes, as well as halibut hooks, digging sticks, bows and fishing floats (Turner and Bell 1971).

ḰÁ,EW̱IȽĆ bark was considered a good medicine. It was boiled together with SȻEṮENIȽĆ (Bitter Cherry) bark to make a cure-all tonic for colds. ḰÁ,EW̱IȽĆ bark was also boiled to make a tea for kidney problems. It was also an ingredient

Pacific Crabapple tree at Island View Beach, near Tsawout.

in Elsie Claxton's special "10-barks" medicine (see pages 30-31). A strip of bark, peeled off in spring, in the morning from the sunrise side of the tree was boiled with the other ingredients. Crabapple bark is said to be good for the appetite; it makes you hungry when you're sick.

Warning: The seeds, bark and leaves of apples, including Pacific crabapple, are potentially toxic if taken in excess. Never take crabapple medicine, or any other medicine, without consulting a knowledgeable herbal specialist.

Note: According to Tim Montler, the domesticated apple (*Malus sylvestris*) is called APELS (7épəls), borrowed from English "apple", and tree is called APELS-IŁĆ (*7epəls-ilhch*).

Black Cottonwood   *Populus balsamifera* subsp. *trichocarpa*
ĆEU,N-EŁĆ, ĆEU,N-EŁP (*chəw'n-əlch, chəw'n-əlhp*)

Black Cottonwood is a tall deciduous tree, growing up to 50 metres high, with resinous, sweet-smelling spring buds and leaves. The bark is smooth and greenish on young trees, and deeply furrowed and greyish on older trees. The leaves are long-stemmed, with heart-shaped blades that are pointed at the tips and finely toothed around the margins. The upper surface is bright green; the lower surface is lighter green. The flowers are long, hanging catkins – males and females are produced on separate trees. At fruiting time in June, the seeds have downy fluff that carries them on the wind; at the peak of their release they look like a summer snowstorm.

Black Cottonwood branch
with a spring bud swollen
with sticky resin.

Black Cottonwood likes damp soil and grows best in marshy areas or along the edges of rivers and lakes. It is common in wet places throughout W̱SÁNEĆ territory, such as Elk Lake and Goldstream.

Traditional Use: Neither Elsie Claxton nor Violet Williams knew the name of cottonwood at first. Violet phoned her sister, Mary Thomas, who said the name, ĆEU,N-EŁĆ. The wood was probably used for fuel and for making the hearth and drill for friction fires. Other First Peoples eat the cambium in the spring, and sometimes make dugout canoes from the logs. The sticky resin from the buds may have been used as a waterproofing and caulking agent, and possibly for making a skin salve as it was by neighbouring First Peoples. But no uses were recorded specifically by the W̱SÁNEĆ people.

Note: This tree is sometimes called ḰEYÁ,LEŚIŁĆ ("dancing-plant"), the same as Trembling Aspen. Cottonwood leaves, too, tend to dance in the wind.

Trembling Aspen *Populus tremuloides*
ḰEYÁ,LEŚIŁĆ, ḰEYÁ,LEŚEŁP (*qw'íy'əl'əsh-ilhch, qw'íy'əl'əsh-əlhp*; meaning "dancing-plant/tree")

Trembling Aspen is a slender tree growing up to 25 metres high. The bark is smooth and greenish white. The leaves have rounded, toothed blades and flat stalks that allow them to shiver and shake with the slightest breeze. The youngest leaves are a delicate yellow-green; they mature to a bluish-green, and finally, in the fall they turn golden yellow before dropping off. Male and female flowers are borne in catkins on separate trees. The seeds are dispersed by the wind and have small parachute-like hairs to carry them. Aspens usually reproduce vegetatively from shoots sprouting up from the roots. They tend to form dense stands, all from the same parent tree.

Trembling Aspen is sporadic but widespread in W̱SÁNEĆ territory. There are small groves at Tsawout and other places around the Saanich Peninsula. It grows in wet sites, often with cottonwood, willows and Red-osier Dogwood.

Traditional Use: At first, Violet Williams called the tree "cottonwood", but later confirmed it as aspen. ḴEYÁ,LEŚIȽĆ is said to be good medicine "for anything". Elsie Claxton said that it is "really good ... for healing anything that's bad in your insides" or any kind of sickness. Simply pour hot water over the bark and drink the infusion.

Aspen bark was an ingredient in Else Claxton's special "ten-barks" medicine (see pages 30-31). A long strip of the bark was used, harvested in the morning from the sunrise side of the tree, and cut into pieces. Some people had heard of the bark being used for birth control.

Note: ḴEYÁ,LEŚIȽĆ means "dancing tree", because the leaves are always moving. Tim Montler points out that it is similar to the name for Lodgepole Pine and White Pine (see page 55). The version that ends in "EȽP" was borrowed from Halkomelem.

Bitter Cherry   *Prunus emarginata*
Tree: SȻETEN̵IȽĆ (*skwt'thəng'-ilhch*)
Bark: DELEM (*t'ə'ləm*)

Bitter Cherry is a small-to-medium-sized deciduous tree with distinctive, smooth reddish-grey bark that is textured with long horizontal lenticels. The leaves are small and elliptical, with fine teeth along the margins. The flowers are small, white and clustered, blooming in spring. The cherries ripen in summer, forming clusters of two to several on a stem. They are small and red when fully ripe. Although they are not poisonous, they taste very bitter, and few people would eat them.

Bitter Cherry grows in open woods, and along roadsides and moist edges of meadows from low to mid elevations throughout W̱SÁNEĆ territory.

Traditional Use: There is apparently no special name for the fruits of this tree, and W̱SÁNEĆ people do not usually eat them. The tough bark, DELEM, was carefully removed from the tree in horizontal strips, in sheets or in spirals cut around the trunk or branches. When scraped with the edge of a knife it becomes smooth and shiny, like a deep-red varnish. It was used for tying and binding implements, including spears, harpoons, arrows and the hafts of bows (Turner and Bell 1971). Christopher Paul said that it was used as a decorative overlay in geometric patterns to cedar-root baskets. Basket weavers from neighbouring First Nations used to dye the bark black by burying it in damp, mucky soil for several weeks; Saanich basket weavers may have done likewise. The black-dyed bark was then used to enhance the patterns in the baskets.

SȻEȾENIⱢĆ wood is said to be a good fuel and people sometimes used it as the hearth and drill in making friction fires.

DELEM also had important medicinal qualities. Some people in the past made a tea from it that was used as a contraceptive, to help space one's children. This tea was also taken for haemorrhaging in childbirth. Elsie Claxton's mother used to use DELEM tea for a sore throat and for tuberculosis. Elsie said that you pour hot water over the bark (four strips) and drink it, just like a tea, and keep drinking it "... and you'll get over that [T.B.]." The bark was also an ingredient in Elsie's special "ten-barks" medicine (see pages 30-31). For this medicine, a strip of cherry bark from the bottom part of the tree was harvested and boiled with the other ingredients. Another "cure-all" medicinal tonic for colds and other ailments was made by boiling DELEM, crabapple bark and other ingredients. W̱SÁNEĆ grandparents bathed their grandchildren in a concoction of SȻEȾENIⱢĆ and wild gooseberry roots to make them intelligent and obedient (Turner and Bell 1971).

Both Violet Williams and Elsie Claxton confirmed that DELEM used for medicine was harvested from the sunrise (southeastern) side of the tree, because the tree is believed to heal faster there, and hence the patient will heal fast, too.

Warning: Bitter Cherry bark, leaves and seeds can be poisonous. Never take medicine without proper knowledge of preparation and dosage.

Bitter Cherry trunk with its bark showing lenticels.

Garry Oak
ĆEN̲ÁȽĆ (*chəng'-élhch*)

Quercus garryana

Garry Oak is a large deciduous tree, growing up to 25 metres tall, with thick, spreading limbs. It often grows in a gnarled and crooked form. The bark is grey, with thickly furrowed ridges. The leaves are shiny and dark green, with deep rounded lobes. The flowers, borne at the tips of the twigs in the spring, are inconspicuous. The male pollen-producing flowers grow in small catkins. The female flowers appear in few-flowered clusters and eventually ripen into brown, capped acorns. Some years produce a heavier acorn crop than others. The leaves turn brown in the fall and then drop to form extensive carpets on the ground.

Garry Oak is restricted in Canada to southwestern British Columbia, mainly on southeastern Vancouver Island and the Gulf Islands. It is common on dry sites throughout W̲SÁNEĆ territory. It grows in two main forms: small, scrubby trees often forming dense groves on rocky hilltops, and large stately spreading individuals in rich-soiled meadows and parklands.

Traditional Use: Long ago, the W̲SÁNEĆ people used to gather and eat acorns, after steaming, roasting or boiling them for a period of time to leach out the bitter tannins, according to Christopher Paul and Homer Barnett (1955). But Elsie Claxton and Violet Williams had never heard of people eating the acorns. Elsie considered them poisonous, even though birds, such as Steller's Jay (ĆIYE), and squirrels enjoy eating them.

ĆEN̲ÁȽĆ bark was an ingredient in various medicinal preparations. Elsie said that a tea made from the bark was good for treating haemorrhaging, and Violet said that it was good for tuberculosis.

A Garry Oak acorn.

Cascara.

Cascara   *Rhamnus purshiana*
KÁYXIȽĆ, or KÁYXEȽP, (*q'éyx̱ilhch, q'éyx̱əlhp*)

Cascara is a small-to-medium-sized deciduous tree growing to about 10 metres tall, with smooth, greyish bark. Its leaves resemble alder leaves, but are smooth around the edges and have prominent veins that run in parallel lines from the middle of the leaf to its edges. The winter leaf buds are also distinctive, because they do not close up tightly, but the miniature developing leaves point outwards at the tips of the twigs. The flowers are small, green and clustered, and the fruits are spherical berries, green at first then ripening to deep, shiny black. They are eaten by Band-tailed Pigeons and many other types of birds, as well as by bears.

Cascara is common throughout W̱SÁNEĆ territory in low and mid elevations. It grows in moist woods with Douglas-fir, Red Alder and Bitter Cherry, among other plants.

Traditional Use: KÁYXIȽĆ was used widely as a laxative. Dave Elliott explains: "The bark was collected in strips in spring or summer and dried. The following summer it was pounded and steeped in cold water. This extract was then boiled. The tonic was drunk as a cure for constipation and other internal ailments." Christopher Paul said that a small piece of the bark soaked in cold water for 12 hours made an excellent tonic. Elsie Claxton said that the bark or branches were put into hot (boiling) water and allowed to steep, like tea. She said that it was bitter and unpleasant to drink.

KÁYXIȽĆ bark was an ingredient in Elsie's special "ten-barks" medicine (see pages 30-31). For this use, a strip of bark, harvested in the morning from the sunrise side of the tree, was cut into pieces and boiled with the other barks.

Note: *Frangula purshiana* is a synonym for *Rhamnus purshiana*. According to Tim Montler, the SENĆOŦEN version of the name ending in "EȽP" is from Halkomelem.

Willows
Hooker's Willow   *Salix hookeriana*
Pacific Willow   *Salix lasiandra*
and others
SX̱ELE,IŁĆ (*sx̱wələ7-íIhch*; from SX̱OLE, meaning "reef net")

Willows are leafy deciduous shrubs or bushy trees, depending on the species. Several kinds grow in W̱SÁNEĆ territory, especially Pacific and Hooker's willows. Apparently, all were called by the same name. (When Mary Thomas was shown a specimen of Scouler's Willow (*Salix scouleriana*), she and Elsie Claxton were uncertain whether this was the "real" SX̱ELE,IŁĆ. Elsie thought the leaves were too wide, but later said it must be a kind of willow.) Probably, the most important type of willow used by the W̱SÁNEĆ is Pacific Willow, which grows as a fair-sized tree with many branches. The bark of the younger branches is smooth and greenish grey; that of the trunk is roughly furrowed and grey. The leaves are lance-shaped, and finely toothed around the edges. In the spring long, crowded catkins are borne, with female and male flowers on separate trees. The female flowers are wind-pollinated, and eventually ripen to release seeds attached to downy parachutes, to be carried away on the wind.

Willows are common and widespread in W̱SÁNEĆ territory. Many, including Pacific Willow, prefer damp places, such as around the edges of lakes or marshes and along creeks or ditches. Others, such as Scouler's Willow, grow in open woods. The best types for making reef-nets come from swamps, such as Maber's Swamp near Tsartlip.

Above: The rounded leaves of Hooker's Willow.

Left: A branch of Pacific Willow showing female catkins.

Traditional Use: Elsie Claxton and Dave Elliott noted that a long time ago, people made reef nets out of the inner bark of willow trees. Dave Elliott said that the bark was processed in the same manner as cedar bark. Dunegrass (*Leymus mollis*) leaves were used to tie around the leaders of these nets, as described below.

Dave told a story about his recollections of the reef net, quoted from a collection of stories he and others told (Elliott et al. 1987):

Learning About the Reef Net

For several summers we visited the family reef-net location on Henry Island. I remember one time when we arrived in the dark, and there were so many salmon jumping that it made a continuous splashing sound, like a turbulent river. I had gone out in a rowboat, and I had seen great schools of salmon passing under the boat. Now I was watching from the shore. They were hauling in the reef net. I wanted to see what was happening, so I moved along the rocky shore as close as I could come to the two anchored canoes. I had heard the watchman in the bow of one of the two 40-foot [12-metre] canoes give the order to haul in. The canoes, each manned by a crew of three, were pulling closer and closer together as the net wrinkled down into the canoes and a silvery torrent of salmon spilled out. Herman Olson walked up to me quite slowly. He was in no rush. He had taken part in the fishing at the family's reef-net location for some years. I came here less often, and I was anxious to find out all I could....

"How do you make the salmon net?" I asked. They were moving salmon to shore in a small scow....

"You fool the salmon," he said....

I asked my mother, "How do you fool the salmon?"

"You make a false bottom, by stringing out approach lines with a special kind of grass [SŁEQÁI, American Dunegrass] attached," my mother said. "Each line is shallower than the one before, so the fish think the water is getting shallower and shallower. They're carried forward with the tide. When they reach the edge of the net, they think they're heading into deep water, so they swim straight down into the net."

"They're heading for the Fraser River, and they run with the tide. We don't have any river in our territory, so our ancestors invented the SX̱OLE [see under Indian Celery, KEXMIN], the reef net, to catch the salmon while they are still in the sea. But it has to be in the right place to catch them. These SX̱OLE locations – they're called SWALETS... – were carefully chosen.

"You know the tree they call the SX̱ELE,IŁĆ," my mother said, giving the ... name for the willow tree. "They used to make cords for the net from the bark stripped from the SX̱ELE,IŁĆ, so they call the net SX̱OLE. The thicker rope for the anchor lines was made of cedar. They use rope from the store now. It takes too long to make willow and cedar ropes. But they still use big rocks for anchors, and cedar logs for floats. You know that."

They were paying out the net again and putting the canoes back on station, swinging them further to the northwest than they had previously been, to face the changing tide.

"I used to go out and get the willow bark for the net," my mother said. "You get the best bark from the big trees in the swamps. I worked here with your JOMEḰ (great grandfather), when he was the boss of this SX̱OLE," my mother said, referring to the great-grandfather in whose house I had stayed in the winter of the big snow....

"The Washington state side is the only place where we can use our SX̱OLE now," my mother said. "They outlawed it on the Canadian side, because they said it was a trap. But they let a fishing company use a real trap, a big trap, at Sooke."

So the SX̱OLE was made out of willow bark from the swampy forest. The cord of the reef net came from the willow tree, and the willow net gave us the salmon. It brought the forest to the sea, and you might say it tied the land and sea together…

Ethnographer Diamond Jenness (n.d.) recorded a story about Sockeye Salmon, part of which explains how the people are taught to make reef-nets from willow:

The Sockeye (and How Willow Bark was Used for a Net)

A few years later Xatstan again brought the sockeye to the Fraser, after they had failed to put in their appearance in the usual numbers; a few only were visible, travelling fast just below the surface of the water. Xatstan ordered some young men to gather him some bark from the willow tree. He made them strip off the outer bark, and to spin the inner fibre into twine. He then taught them to make a bag net (*sahe'in* = Saanich *swaltam*) with the twine, and to tie two rocks on each side as sinkers. Then he sent them out to fish with it, dragging it downstream from two canoes. They caught two sockeye, one male and one female....

Homer Barnett (1955) reported that willow bark was also used to make fishing lines and for basket decoration. Willow charcoal is said to be good for protective face paint, like that of devil's-club. The wood is cut into little pieces and burned, then the charcoal is rubbed on the face.

Violet Williams had heard that willow bark was supposed to be good for fevers, although she had never taken it herself. Willow bark was an ingredient in Elsie Claxton's special medicine (see pages 30-31). "Pussy Willow" (Pacific Willow), a big tree with pussy willows in the spring, was the type normally used in this preparation. A narrow strip of bark was harvested in the morning from the sunrise side of the tree, cut into pieces, and boiled with the other ingredients.

Shrubs and Woody Vines

Saskatoon
Amelanchier alnifolia
Berries: SĆI,SE̱N (schí7səng)
Bush: SĆI,SE̱N I̱ŁĆ (schi7sən-ílhch)
Other names: Serviceberry, Saskatoon Serviceberry, Saskatoonberry.

Saskatoon grows as a medium to high deciduous shrub, up to three metres tall. The smooth stems are reddish brown to dark grey and the young twigs often silky. The leaves are oval and up to three centimetres long. Sharp teeth line the upper half or two-thirds of the leaf margins. The young leaves are bright green but turn bluish green with age. Showy white flowers in dense clusters appear in late May and June. By early July, the berries ripen. They are spherical, up to a centimetre across, and range from dark blue to nearly black. They are covered by a greyish-blue waxy coating. When fully developed the fruit is juicy and sweet. But on the coast it often dries out quickly, becoming mealy or crunchy with little flavour. The leaves turn yellow in the fall before dropping off.

This shrub favours open to lightly shaded sites, such as thickets, fence rows, clearings and edges of woods. Sometimes it grows along the rocky coastline. Saskatoon is common but sporadic throughout W̱SÁNEĆ territory.

Traditional Use: People eat the berries fresh in the summertime. They seldom dried them, although Elsie Claxton said that her mother used to make jam from SĆI,SE̱N. They were one of Vi Williams' favourite fruits. Elsie said that there used to be lots of SĆI,SE̱N around, but now they are uncommon even where the bushes are plentiful.

The W̱SÁNEĆ people sometimes used SĆI,SE̱N IȽĆ wood for arrow shafts (Jenness 1945). It is very hard, and the branches can grow very straight.

SĆI,SE̱N IȽĆ bark was an ingredient in Elsie Claxton's special "ten-barks" medicine (see pages 30-31). A strip of bark, harvested in the morning from the sunrise side of the bush, was cut into pieces and boiled with the other ingredients.

Red-osier Dogwood
Cornus stolonifera
NEȻIM S̱X̱EL,I,EȽĆ (*nəkwim sx̱wələ7ilhch*)
Other name: *Cornus sericea.*

Red-osier Dogwood is a medium to tall deciduous shrub with smooth-barked stems that are often wine-red, or sometimes yellowish green. It can grow very tall, but usually to about five metres. The stems often arch and root at the ground, forming dense thickets. The branches grow in opposite pairs, each pair positioned at right angles to the pair above and below. The leaves are oval to elliptical, pointed and usually about nine centimetres long. The upper surface is medium to dark green and the underside distinctly grey-green. In the fall, the leaves usually

turn maroon before falling. The flowers are small and creamy, growing in dense, round-topped clusters. Waxy white to light-blue berries develop in the late summer and often persist into late fall; they have a hard stone in the middle and they taste bitter but are not poisonous.

Red-osier Dogwood ranges from low to moderately high elevations throughout W̱SÁNEĆ territory. It prefers moist sites in open woods, in swamps and at the edges of watercourses, where it can grow very large.

Traditional Use: Elsie Claxton and Violet Williams both recognized this shrub, and commented that they used to harvest the branches to make knitting needles. They didn't remember the name, but Dr Earl Claxton Sr remembered it and wrote it out. Another shrub whose name was not remembered, but whose wood was also used for knitting needles, is Pacific Ninebark (*Physocarpus capitatus*). Elsie Claxton noted that when the branches of ninebark are bigger the wood is nicely white. Ninebark grows in moist swamps and along ditches, and is often seen together with Red-osier Dogwood.

As a medicine, NEȻIM S̱X̱EL,I,EȽĆ bark soaked in warm water produces a solution that induces vomiting. It was said to clean out the stomach and improve breathing. This medicine was used by canoe-pullers before races and by others undergoing purification.

Note: Red-osier Dogwood is sometimes called Red Willow, although it is a true dogwood, related to the Flowering Dogwood, BC's provincial emblem. The W̱SÁNEĆ name, NEȻIM SX̱EL,I,EȽĆ means "red willow".

Hazelnut   *Corylus cornuta*
Nuts: ḴEBOX̱ (*qwp'áx̱w*)
Bushes: ḴEBOX̱ IȽĆ (*qwp'ax̱w-ílhch*)

Hazelnut is a large, deciduous shrub, growing up to four metres tall, with many stems. The young twigs and leaves are covered in dense white hairs. The leaves are broadly oval to elliptical and are borne alternately along the twigs. They commonly have a heart-shaped base and a pointed tip, with serrated edges. Male flowers grow in catkins, which hang down from the branches before the leaves appear in the spring. The female flowers are a beautiful deep maroon, but very small and inconspicuous. Spherical edible nuts are produced inside tubular husks. The husks are light green and covered in stiff, sharp hairs. The husks are usually clustered in twos or threes at the ends of the branches.

Hazelnut has a restricted range on Vancouver Island, but occurs in W̱SÁNEĆ territory around Goldstream and on the slopes of Malahat Ridge in particular. It prefers moist, well drained sites at low to middle elevations. It frequents open forest, shady openings, thickets, clearings, rocky slopes and stream banks.

Traditional Use: Elsie Claxton used to gather ḴEBOX̱ from Puyallup, Washington when she was a young teenager. Violet Williams knew of ḴEBOX̱ from a lady at Coles Bay who grew the domesticated type. At Puyallup, Elsie Claxton and her family picked the nuts, took off the prickly covering, spread them on cookie sheets and roasted them in an oven. They ate them still warm from the oven. The nuts are still growing on the reserve at Pauquachin. Dave Elliott noted, "the nuts were a favourite food of our people and were gathered at Goldstream." Squirrels and other wildlife also enjoy the nuts.

Christopher Paul, who remembered gathering ḴEBOX̱ at Goldstream, also noted that the straight suckers of the ḴEBOX̱ IȽĆ were sometimes used to make arrow shafts by W̱SÁNEĆ hunters.

Black Hawthorn   *Crataegus douglasii*
MÁȾEN IŁĆ (*meyt'thən7ilhch*)

Black Hawthorn is a large, bushy deciduous shrub or small tree with greyish bark and tough, spreading leafy branches. Mature plants can reach four metres

or higher. The branches are armed with stout, sharp spines, usually a centimetre or two long. The leaves are thick, dark green and shiny, roughly oval and coarsely toothed around the top. The flowers bloom around May in white, showy, rounded clusters. The fruits are clustered, shiny black and seedy.

Black Hawthorn is common in moist thickets and edges of woods and meadows from low to mid elevations. It occurs throughout W̱SÁNEĆ territory, especially along the coast.

Traditional Use: Mary Thomas recalled that when she was little she once climbed a MÁȾEN IŁĆ and ate the berries, and gave some to her younger siblings as well. Her mother came and scolded her, and told her that those berries were "crow's berries" not to be eaten by people. The berries are not actually poisonous, but they tend to be dry and seedy. Some First Nations in the interior eat them and dry them in cakes for winter. There are some reports that the Vancouver Island Salish peoples also ate them, but with salmon roe because they were dry.

The wood is very tough, and some First Peoples use it for digging sticks and other implements needing strength. It is a ritually important plant as well, because of its sharp spines. Dancers sometimes use the charcoal for face paint, as they do with Devil's Club charcoal.

Salal   *Gaultheria shallon*
Berries: DAḴE, (*t'eqə7*)
Bush: DAḴE IŁĆ (*t'qe7-ilhch*)

Salal is an evergreen shrub, growing up to two metres tall on the rainy west coast, but usually one metre or less in W̱SÁNEĆ territory on the eastern side of Vancouver Island and the Gulf Islands. Its branching stems are tough and wiry, and are densely covered by numerous thick, leathery, oval, pointed leaves with finely toothed margins. The flowers are white to pinkish, urn-shaped and arranged in a row of several to many along the stalk, all the flowers pointing downwards. The fruits are dark purplish to blackish juicy, sweet berries with many tiny seeds. They are slightly hairy, and at the bottom is a star-shaped depression.

Salal grows in partially shaded coniferous and mixed forests at low to mid elevations. It often forms dense, almost impenetrable thickets with its tough, wiry branches. It is common throughout W̱SÁNEĆ territory, except in the driest places.

Traditional Use: Many people still pick these berries. Elsie Claxton and Violet Williams really enjoyed picking them. Elsie said that there used to be large, juicy DAḴE around the Tsawout Reserve, but in recent years they are all small and dried up, and not nearly as good as they used to be. In the past, people cooked the berries and spread them out on the rocks or on skunk-cabbage leaves to dry in the sun or over a fire. Sometimes the berries were dried individually, like raisins. More recently, people jar them as preserves, freeze them, or make them into jam or jelly. Violet's mother used to just cook them up and serve them. Elsie's mother used to dry all her berries, even soapberries. She stored them for winter in boxes, baskets or bags. Then, before serving them, she soaked them in water until they were just like fresh berries.

In pit-cooking camas bulbs and clams, DAḴE IŁĆ branches and fern fronds were used to protect and flavour the food in the steaming pits. DAḴE IŁĆ branches were also bunched together and used as a whisk for whipping up Soapberries.

DAḴE IŁĆ leaves are said to be good medicine. Once when Elsie's brother, Sandy Pelkey, was hunting in the mountains, he saw a deer that had been shot on its hip. It was chewing DAḴE IŁĆ leaves and spitting them onto its wound, and the wound appeared to be healing well. This was an indication that DAḴE IŁĆ leaves would be a good medicine to use on cuts and wounds.

It is said that Octopus chews DAḴE with pitch to make his ink.

Note: Nowadays, Salal branches with their handsome dark-green leaves are widely sought by florists for use in floral bouquets and flower arrangements. If they are kept moist and cool, they will remain fresh for many weeks. Salal from Vancouver Island is shipped all over the world.

Oceanspray
KÁȾEȽĆ (q'éy'ťth-əlhch)

Holodiscus discolor

Oceanspray is a tall, many-stemmed deciduous shrub from 1.5 to 5 metres high. Clumps reach from 2 metres to nearly 10 metres across, sometimes forming dense stands. Several stems rise upward from the centre of the clump. Where large branches begin to bend, vigorous young shoots arise. These straight shoots are the most useful part of the shrub. The bark of young shoots is brown and sometimes slightly angled and striated. The older bark is light grey and nearly smooth. Light green, lobed, oval, soft-textured leaves about 4-10 cm long scatter along the upper half of the branches. Small to large teeth line the margins of the leaves. Plumes

of creamy white flowers cascade from the tips of the branches in late June and early July, lasting two or three weeks. Each 10-20-cm-long cluster contains hundreds of tiny whitish flowers. Great quantities of pollen attract hordes of pollinating insects. The blossoms release an exotic creamy scent, almost overpowering within a dense thicket. Once the flowers fade, the clusters turn dry brown.

Oceanspray favours sunny to slightly shaded places in open woods and on the edges of clearings. It thrives in shallow rocky soils with good drainage. Typically it will form a thicket on a rocky bluff at the edge of a wooded area and in an open forest. It grows along the coast at low to mid elevations.

Traditional Use: Elsie Claxton used to say that when KÁȾEȽĆ was in full bloom, it was the best time to start fishing for Sockeye Salmon. KÁȾEȽĆ stems grow very straight and become hard after heating them. Because of this quality, the shrub is sometimes called "ironwood". It was prized by the W̱SÁNEĆ for making bows, arrows, camas-bulb digging sticks, harpoon shafts, salmon barbecuing sticks, cambium scrapers, halibut hooks, cattail-mat needles, knitting needles and many other items, according to Christopher Paul and John Elliott. Elsie's son, Earl, once made a halibut hook from KÁȾEȽĆ wood.

After clams were pit-cooked in the traditional way, they were taken out of the shells, and skewered on KÁȾEȽĆ sticks about a metre long. These were poked into the ground above a fire down at the beach and the clams were allowed to cook over the coals, being turned as needed; they should not be cooked too long. Once barbecued like this, the clams were dried, to be stored for winter or traded with people at Yakima and elsewhere, who really craved this coastal delicacy.

The brown fruiting clusters of KÁȾEŁĆ are good for treating diarrhea. An infusion was prepared by pouring hot water over the clusters after they had been cleaned. People used to take this medicine when they went to Yakima to pick hops, apples and grapes. The coast people, especially children, often used to get diarrhea when they went to Yakima. Many children got sick and died over there. The W̱SÁNEĆ people were told by the Yakima that the flies had been bad there ever since the Nez Perce Indian War of 1877 (a war between the Nez Perce Nation in Washington and the United States government), when there had been a lot of unburied bodies. The flies arrived at that time, and are said to have been there ever since. The flies are believed to be the cause of much sickness with diarrhea and other illnesses.

Orange Honeysuckle
Lonicera ciliosa

KIDE,A̱N EŁP (*q'ít'ə7əy'n-əlhp*) or KIDE (*q'ít'ə*; meaning "swing") or KIDE EȾ SPELḰIȾE (*q'ít'ə 7ə cə spəlqwít'thə7*; meaning "swing of the screech owl/ghost")
Other name: Western Trumpet.

Orange Honeysuckle is a branching vine twining its way up shrubs and trees to heights of many metres. Stringy greyish bark covers the flexible stems. Once the vine reaches sun, the stem divides many times to form a cascading mantle over the supporting tree or shrub. The oppositely paired, bright green leaves are elliptical and smooth around the margins. Most distinctive are the terminal leaves surrounding the flower cluster. They are joined at the base to form an oval saucer-like structure beneath the tight cluster of bright-orange, tubular flowers. There may be as few as one or two blooms or as many as thirty in a cluster, depending upon the vigour of the plant. Near the base of each flower tube there is a pouch that collects sweet nectar, an offering for the pollinating hummingbirds. Flowering time is late spring and early summer. Bright red or orange juicy berries replace the fertilized flowers. These look appetizing, but are not edible.

Orange Honeysuckle grows in woods and thickets, and also grows abundantly on shrubs and trees along roadsides. It is common throughout W̱SÁNEĆ territory. A smaller, purple-flowered species, Hairy Honeysuckle (*Lonicera hispidula*) also commonly grows in the region. It usually does not climb up trees and high shrubs but seems to stay close to the ground.

Traditional Use: Mary Thomas reported that KIDE‚ÁN EȽP is a very strong love medicine. But it's said to be so strong that if your spouse or partner dies, you will die too. The medicine makes you want to cling to someone just like the vine clings around the tree. The plant is simply called "honeysuckle" by people today. The original name originates from the way the vines swing by themselves in the wind.

The nectar-rich flowers are still enjoyed as a treat by young children, who bite into the nectar pouch and suck out the sweet fluid.

Tall Oregon-grape
Dull-leaved Oregon-grape
Berries: SENI, (səní7)
Plant: SENI, IȽĆ (səni7-ílhch)

Mahonia aquifolium
Mahonia nervosa

Tall Oregon-grape is a spindly shrub with upright greyish canes. It can grow to more than two metres tall, but in dry sites it is much shorter. The inner bark of the stems and roots is bright yellow. The leaves are evergreen, shiny and compound, each with five to nine leaflets. Spiny teeth line the edges of the leaflets. In the summer, the leaves are lustrous green, but in the winter they often turn bronze or red. The new growth in the spring is a reddish colour as well. The flowers are small and bright yellow, and form large showy clusters in spring. Grape-like bunches of spherical blue, sour-tasting berries replace the flowers in the summer.

A related species, Dull-leaved Oregon-grape, is similar but is usually shorter and has leaflets that are usually narrower and more numerous per leaf. As the name implies, its leaves are not as shiny as those of Tall Oregon-grape. Its berries are borne in elongated clusters.

Tall Oregon-grape grows in open woodlands, along the edges of meadows and on partially shaded rocky outcroppings. Dull-leaved Oregon grape usually grows in shaded woods. Both are common throughout W̱SÁNEĆ territory. Oregon-grape is a common ornamental plant and can be purchased in most nurseries.

Tall Oregon-grape.

Traditional Use: Both species of Oregon-grape are called by the same name and are used in similar ways. Elsie Claxton's mother told her that in the old days, the only antidote for shellfish poisoning was to eat large quantities of fresh, sour SENI,. They are said to be good for any other kind of poisoning as well. The tart berries were also eaten raw or boiled as a regular food, alone or mixed with other berries, like Salal, and are commonly made into jelly.

The bark of the roots and stems – especially the inner bark – used to be shredded and boiled in water to make a yellow dye for basketry and for colouring wool.

Dave Elliott noted that people pounded and boiled the stems and roots, then drank the extract as a remedy for skin diseases, and as a general tonic to induce a reviving feeling. The root extract, he said, also made an excellent detergent lotion.

Elsie Claxton holding branches of Dull-leaved Oregon-grape.

June Plum
Oemleria cerasiformis

Fruit: ȾEX̱EN (*t'thə'x̱wən'*)
Bush: ȾEX̱EN IȽĆ (*t'thəx̱wən'-íłhch*)
Other names: Indian Plum, Bird-cherry.

June Plum is a tall, deciduous shrub with oblong, smooth-edged leaves, and a pungent odour to its bark and twigs. The plant produces leaves and flowers earlier than any other shrub in the area. The small white flowers are borne in elongated drooping clusters, with male and female flowers usually on separate bushes. By

late June and early July the female flowers ripen into clusters of small dark blue plum-like fruits that are pleasant tasting, but have thin flesh surrounding a large stone.

June Plum is widespread in W̱SÁNEĆ territory, growing in moist open woods and roadside thickets.

Traditional Use: Elsie Claxton and Dave Elliott noted that ȾEXEN IȽĆ is the first bush to leaf out in the spring. The buds start to burst open with bright emerald-green leaves as early as February. Christopher Paul, Violet Williams and Elsie Claxton used to eat the small, grape-like berries as kids.

Elsie Claxton recalled that the bark of this shrub was a good medicine for diarrhea and as a purgative to "clean all bad things". She said it makes you vomit, and "takes out everything". You would take it as a medicine whenever you needed cleansing, not just in the spring.

The sticks of ȾEXEN IȽĆ were an ingredient in Elsie Claxton's special "ten-barks" medicine (see pages 30-31). Sticks of about 30 cm long were cut into pieces and boiled with the other ingredients.

Devil's Club
ḴO,PEȽĆ (qwá7p-əlhch)
Oplopanax horridus

Devil's Club is an erect to sprawling spiny shrub that grows up to three metres tall, with thick, often tangled light-grey stems. The wood has a sweet odour. The large leaves are shaped like maple leaves; they alternate along the stem on long

stalks. Each leaf has seven to nine pointed, toothed lobes, and the undersides are spiny along the veins. The small whitish flowers grow in a dense pyramid-shaped cluster at the top of the shrub. They ripen into bright red, strong-smelling berries.

Devil's Club is not known to occur on the Saanich Peninsula, but is found in the mountains of the Malahat and Sooke Hills. It is very common on the wetter parts of Vancouver Island and the mainland. It grows in moist, shaded woods, in wet seepage areas, along streams and in avalanche tracks at low to mid elevations.

Traditional Use: Elsie Claxton said that K̇O,PEȽĊ spines are poisonous, and that you should not touch them. She said she would never drink or eat something containing K̇O,PEȽĊ, because she believed that the entire plant was poisonous. But Violet Williams thought that K̇O,PEȽĊ was used to make a medicinal tea for diabetes. She explained that the branches were cut into pieces and boiled with the bark of BEN,Á,YEȽP (Douglas Maple). She knew a Quw'utsun' woman who was told she had diabetes and that she would have to take insulin injections every day. She decided instead that she would try taking traditional medicine, the K̇O,PEȽĊ tea, and when she went back to the doctor for another blood test, the symptoms had disappeared. Violet added that many aboriginal people have died from diabetes.

Dave Elliott said, "The roots were pounded, boiled and used as a poultice for rheumatism and other aches. The prickly stems were beaten against the skin for sore limbs."

Elsie Claxton, Violet Williams and Dave Elliott all recalled that the charcoal from this plant was used as a (ceremonial) face paint for dancers. It was powdered and mixed with grease. It was also used for a bluish tattoo (Jenness 1945).

Bog Cranberry *Oxycoccus oxycoccos*
Vines K̇EMĊOLS IȽĊ (*qwəm'chal's-ílhch*)
Berries K̇EMĊOLES (*qwəm'chálǝs*)
Other name: *Vaccinium oxycoccos*.

Bog Cranberries grow on tiny vines half hidden in the moss. The stems are wire-like and the leaves are small and oval shaped. The flowers are small and pink, with recurving petals, and the berries are borne on thin stalks. They ripen in the fall. When they are still green but full sized they can be picked and preserved. Bog Cranberries are related to the cranberries grown in the Fraser Delta area, from which cranberry juice and cranberry sauce are prepared.

Looking eastward at Maber's flats, flooded in winter.

Bog Cranberries grow in acidic peat bogs with Labrador tea, Lodgepole Pine and sphagnum mosses. They are hard to find.

Traditional Use: Elsie Claxton was told by Henry Smith that people used to go to Royal Oak (i.e., Rithet's Bog) for K̲EMĆOLES, although she had never been there herself. (She also stated that had never heard of people gathering any kind of "tea" plant from this site, only cranberries.) Christopher Paul reported that people long ago picked K̲EMĆOLES at Rithet's Bog. Violet Williams had heard that they could be found somewhere on the west coast, possibly in the muskeg areas above Jordan River. Vi and Elsie said that people used to go over to the swamp in the Fraser River valley (probably Burns Bog), where lots of these berries grew. They also used to grow at the end of Langford Lake, and, long ago, before Maber's Swamp was drained, the W̲SÁNEĆ people used to get K̲EMĆOLES there in abundance. The berries are quite tart. They could be eaten fresh, or were stored under water in containers.

Stink Currant
Berries: SPEⱮ (*spət'th*)
Bush: SPEⱮIŁĆ (*spət'thilhch*)

Ribes bracteosum

Stink Currant is a lanky deciduous shrub with a distinctive rank smell to the bark and leaves. It grows up to three metres high, with slender, grey-barked stems, spotted with yellowish glands. Its leaves are large and maple-like, with 5-7 lobes. They closely resemble thimbleberry leaves, but have more pointed lobes. The flowers are small and numerous, greenish or whitish and growing in long clusters. The berries are spherical and dark purple in colour, but dusted with a whitish waxy coating that renders them light blue. They are spotted with small glands, and, like other currants, have a brownish wick at the end, remnants of the flower.

Stink currant grows in moist, shaded areas, along creeks and in forested swamps. It is common at Goldstream.

Traditional Use: Elsie Claxton and Violet Williams recognized this shrub, and knew that its berries were edible, but they had not eaten them. They knew of no particular use for the plant, but knew its name.

Domesticated garden currants are called KELENTS (*kə'lənts*; borrowed from English "currants").

Notes: Many First Peoples on the coast prefer Greyberry as the English common name for *Ribes bracteosum*. The W̱SÁNEĆ name for the berries, SPEⱮ, is similar to PEⱮ (*pət'th*), which means "sour, mousey smell, like a tomcat or skunk". According to Elsie, this name is close to but not the same as SPA,EⱮ (*spe7əth*), the word for Black Bear.

Wild Black Gooseberry
Berries: K̲ÁMQ (qémkw')
Bush: K̲ÁMQ IŁĆ (qémkw'-ílhch)

Ribes divaricatum

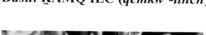

Wild Black Gooseberry is a stout, thorny deciduous shrub with small, maple-like leaves and one to three sharp spines at each node. It grows up to three metres tall. Its flowers are drooping, and have deep red sepals, white petals and protruding stamens. The fruits are smooth and purplish black when ripe. They ripen in midsummer and have a distinctive, tart taste.

This shrub occurs throughout W̱SÁNEĆ territory, often along the coast, but it is not common.

Traditional Use: According to Dave Elliott, K̲ÁMQ were usually boiled and then dried into cakes. They were often mixed with Salal berries.

K̲ÁMQ IŁĆ roots were apparently boiled with cedar roots and rose roots, pounded and twisted into rope. They were also used to help construct the W̱SÁNEĆ reef nets. Willow bark was the main material for these.

Dave Elliott noted, "The sharp thorns were used as probes for boils and removing splinters. They were also used for tattooing. Boiled roots were rubbed on the skin for severe cramps (charley horse) and other ailments." Grandparents, he said, used to wash their newborn grandchildren in a solution of K̲ÁMQ IŁĆ and Bitter Cherry roots so they would grow up to be intelligent and obedient.

Note: Gooseberry Point on Orcas Island is called DÁMW̱ĆIKSEN (from the Halkomelem words for "gooseberry" and "point", according to Dave Elliott and Dr Earl Claxton Sr). According to Tim Montler, the Saanich equivalent would probably be K̲ÁMQIK̲SEN. Another type of gooseberry was apparently called by the same name: Gummy Gooseberry (*Ribes lobbii*) is a relative that has larger deep-red flowers and large, reddish, somewhat fuzzy berries.

Red-flowering Currant
Berries: W̱IW̱Q, (*xwíxwkw'*)
Bush: W̱IW̱Q,IȽĆ (*xwixwkw'-ílhch*)

Ribes sanguineum

Red-flowering Currant is a deciduous bushy shrub, growing upright to three metres tall. The bark is greyish and the leaves grow in clusters on spur branches spaced along the stems. Each greyish-green leaf looks like a small rounded maple leaf. These currants bloom in the early spring, usually around March and April. At this time, they can be completely covered with drooping clusters of brilliant deep pink (occasionally pale pink or white) blossoms. The berries grow in clusters, ripen in early summer and appear pale blue from a whitish waxy coating. They are somewhat seedy and dry, but are enjoyed by many kinds of birds, as well as people. Hummingbirds seek the flower nectar of these bright blooms, which are among the first available hummingbird flowers of the season.

Red-flowering Currant grows in dry open woodlands, on rocky slopes, along roadsides and at the edges of clearings at low to middle elevations. It is also commonly grown as an ornamental in peoples' gardens.

Traditional Use: Elsie Claxton knew this shrub as one of the first to flower in spring. She liked it very much, but had not heard of the berries being eaten. But people would have eaten W̱IW̱Q in the past, and probably also dried them into cakes for winter use. Likely they were mixed with Salal berries and other types of berries.

Mary Thomas called the bush "wild lilac." Their parents used to call it "rain flower", and she noted that "if you break them, it rains a lot". She recalled that as a child, she and her friends got into trouble for breaking off the flowers and causing it to rain.

Nootka Rose   *Rosa nutkana*
Flowers: K̲ELK̲ (*qél'q*)
Fruit or hips: K̲ILEK̲ (*qə'l'əq*)
Bush: K̲ELK̲E IⱢĆ (*qəl'əq-íⱡhch*)
Other name: Common Wild Rose.

Wild Rose is a deciduous shrub that varies in form from tall and bushy to slender and somewhat spindly, depending on the amount of sunlight and moisture it receives. Sometimes it forms dense, impenetrable thickets along roadsides and hedgerows. It can grow to heights of three metres. The branches bear stout thorns at the leaf nodes, as well as smaller thorns along the length. The leaves are compound, with five to seven oval, toothed leaflets attached to a central stalk. In some forms, the foliage is strongly scented: one of the most delicious perfumes in nature. Warm, late spring sun coaxes forth the aromatic scent from thickets along the sea shore. Pink five-petalled blossoms, some as wide as five centimetres, adorn the thorny thickets in late spring. The flowers usually occur singly. In late summer they mature into spherical orange-red fruits, called "hips". The fleshy rind of the hips surrounds whitish seeds covered in tiny hooked hairs.

Nootka rose is common throughout W̲SÁNEĆ territory. This thorny shrub forms dense thickets, together with other species such as Common Snowberry, June Plum and Red-osier Dogwood. It also grows in open areas and in light woods. Damp but not wet sites bring out spectacular growth, especially in warm, dry regions.

Two other species of wild roses occur in W̲SÁNEĆ territory: Swamp Rose (*Rosa pisocarpa*), which grows in damp places and has smaller, more clustered flowers; and Dwarf Rose (*R. gymnocarpa*), which grows in shaded Douglas-fir woods and is smaller and more delicate, with numerous fine spines along the greenish stems. It has tiny pink blossoms and small ovoid hips that do not have the leafy sepals attached at the ends.

Nootka Rose flowers, hips and (on facing page) thorns.

A close look at a tiny Dwarf Rose flower.

Traditional Use: Christopher Paul said that the tender young shoots of ḴELḴE IŁĆ were sometimes peeled and eaten in the spring, and that the ḴILEḴ were eaten raw in the autumn, after they had been touched by frost and turned red. (But never eat the seeds inside the hips, because they are covered with stiff hairs that can cause irritation, just like itching powder.)

ḴELḴE IŁĆ roots were peeled and boiled, then used with gooseberry and cedar roots in the making of reef nets.

ḴELḴE IŁĆ branches were broken off and boiled to make an eye medicine, the solution used to flush the eyes of someone not able to see well. In the old days, this solution was used to treat any eye problem, including cataracts.

Swamp Rose would have been named and used similarly to Wild Rose. Dwarf Rose, perhaps because of its prickly quality, is a highly spiritual plant. It has a special name, ḴEMI,EŁĆ (*qəmi7-ílhch*), and was used for bringing strength and ritual protection for young people at puberty, new dancers, and recently bereaved people who had lost parents, a spouse or a close relative. Violet Williams said that the stem of this prickly shrub was also used to clean the inside of gun barrels. People sometimes ate the hips of this small rose, according to Wayne Suttles (1951).

Thickets of Nootka Rose and other shrubs on the southern bluffs of Beacon Hill Park, Victoria.

Blackcap   *Rubus leucodermis*
Berries: NEKIXTKOME (*nəq'íx̲ t'thqw'am'ə7***; meaning "black blackcap")**
Bush: TKOME,IŁĆ (*t'thqw'am'ə7-ílhch***)**
Other name: Black Raspberry.

Blackcap is a raspberry-like plant that has arching bluish canes. It can grow roots and start new plants where the tips of the canes touch the ground. The leaves are three-parted, like raspberries, the flowers are white, and the berries are similar to raspberries, but finer grained and dark purplish blue in colour when ripe. They are sweet and juicy.

Blackcaps prefer open places at the edges of woods, along old roads and even in clearcut areas, where they can produce fruit prolifically in a good year.

Traditional Use: These sweet, juicy berries were dried and stored for winter use. They were mashed and packed in rectangular frames and dried in the sun. The dried NEKIXTKOME cakes were two or three centimetres thick and cut into squares of about thirty centimetres. They were soaked in water overnight to rehydrate them for eating. These soaked berries tasted just like fresh ones, according to Chris Paul.

NEKIXTKOME are plentiful on Malahat Ridge. W̱SÁNEĆ people really like them. Old timers used to go and get them a long time ago, probably by canoe; now people go by car to get them.

Note: The Quw'utsun' call Red Raspberry (*Rubus idaeus*), which is not native to Vancouver Island, *t'thqw'am'ə7in* – the same as the name for Blackcaps, but without the "black" modifier. The SENĆOŦEN name for Red Raspberries is TKÁN̲E, (*t'thqw'ang'ə7*). According to Tim Montler, the SENĆOŦEN name for Blackcap itself, is borrowed from Quw'utsun', as indicated by the *m'*, rather than *ng'*.

Thimbleberry   *Rubus parviflorus*
Berries: DEK̲EN̲ (*t'ə´qwəng*)
Bush: DEK̲EN̲ IȽĆ (*t'əqwəng-iȽhch*)
Edible sprouts: ȾAȾKI (*thé7thq'i*)

Thimbleberry is a deciduous shrub with large maple-like leaves that forms dense patches. Unlike the closely related raspberries and blackberries, Thimbleberry stems have no thorns. The skin on the young shoots is fuzzy, and the bark of the older stems is light brown and shredded. The leaves are soft and somewhat fuzzy, with long stalks and blades usually 15-20 cm across. The leaves have seven to nine pointed lobes. The flowers are white and showy, borne in clusters of 3 to 11 blooms. Flowering occurs in late spring, and the flowers can reach up to 5 cm across. The berries ripen in July through August, depending on elevation. They are shaped like shallow thimbles. When ripe they are bright red, soft and very sweet. They are especially good if they have been allowed to ripen in sunshine with plenty of moisture around their roots. The leaves turn yellow in the fall and drop off, leaving numerous erect, brown-barked stems, barely noticeable in winter thickets.

Thimbleberry is common and widespread, growing in open sites, often at the edges of woods, roadsides and shorelines. It can be found in both moist and dry sites from sea level to high elevations.

Traditional Use: Christopher Paul said that the sweet, juicy ȾAȾKI were harvested, peeled and eaten raw in the spring. They had to be harvested before they turned woody; if you could snap the stalk off with your fingers, it was just right to peel and eat. Violet Williams recalled eating these sprouts when she was a girl.

DEK̲EN̲ were picked soft and sweet wherever they were found, and eaten fresh, boiled or pressed and dried into cakes for winter use.

As an effective medicine for diarrhoea, and to ease a stomach-ache, dried brown DEK̲EN̲ IȽĆ leaves could be chewed, or a tea made from the leaves, according to Chris Paul, Dave Elliott and Violet Williams.

Salmonberry   *Rubus spectabilis*
Berries: ELILE (ʔəlílə ʔ, or lílə ʔ)
Bush: ELILE IȽĆ (ʔəlílə ʔ-ílhch)
Edible sprouts: ȽÁ,ȽKI (thé ʔthq'i)

The W̱SÁNEĆ people recognize these colour forms of the berries:
  PELPEKXELIꞰ (*pəlpəq'xəlíqw*) – very light ("little white heads";
      cf. PEK (*pəq'*) "white")
  NEKIX (*nəq'íx̱*, meaning "black") – dark red
  NENELPW̱IꞰ (*nənəl'pxwíqw*, meaning "little brown heads")
      or NELPW̱IꞰ (*nəlpxwíqw*; cf. NEPEW̱ (*nəpə'xw*), meaning "blond,
      yellowish-brown") – golden
  NENELꞰEMEꞰ (*nənəl'kwəmíqw*, meaning "little red heads";
      cf. NEꞰIM (*nəkwím*, meaning "red")) – ruby

Salmonberry is a raspberry-like upright deciduous shrub growing up to three
metres tall. Its bark is light tan to reddish brown and it has prickles along the
stems. The leaves are like raspberry leaves: three-parted, each leaflet being pointed and sharply toothed along the margins. The flowers bloom early in the spring,

often before the leaves have expanded.
They are pale to deep pink. The berries
resemble raspberries but are larger and
coarser; they vary from light orange to
deep ruby or almost black. Each bush
has its own distinctive colour of berry;
most bushes that bear ruby berries also
have reddish bark and a reddish tinge
to their leaves.

Salmonberry grows in moist forests
and edges of creeks, lakes and swamps,
where it often forms dense thickets. It
grows from low to mid elevations.

Traditional Use: Dave Elliott noted that ȽÁ,ȽKI usually provided the first fresh
food of the early spring. They were peeled and eaten raw, according to Christopher Paul, or steamed with salmon or salmon roe. Violet Williams and Elsie
Claxton recalled eating the sprouts when they were young, and really enjoyed
them. There are many ELILE IȽĆ at Goldstream, especially NENELPW̱IꞰ, the
golden-berry type. When we went there with Elsie and Vi, we heard the Swainson's Thrush singing, to make the berries ripen (see story, below).

People generally ate ELILE fresh and raw; they were usually considered too
juicy to dry into cakes, according to Dave Elliott. They are still a favourite food –
it's hard to beat plump, juicy, fully ripe ELILE as a dessert.

Shredded, boiled salmonberry bark is said to be an effective poultice for cuts and wounds. You can also soak a cloth in a solution made by boiling the stems, and use it as a compress on a wound. Elsie Claxton's mother used to make a tea from cut-up ELILE IŁĆ stems to treat diarrhoea. She would allow the solution to cool before using it.

Hummingbirds, warblers and other birds seek out the sweet nectar from ELILE IŁĆ blossoms in the spring, and the fruits are eaten by bears, robins and many other types of wildlife.

Swainson's Thrush, widely known as the Salmonberry bird, is called W̱EW̱ELES̱ (*xwəxwə'ləsh*). When it sings "W̱EW̱ELEW̱ELEW̱ELEW̱ES̱!" (*xwəxwələxwələxwəlxwəsh*!) in rising flute-like notes the ELILE get their colour and ripen. Violet Williams and Elsie Claxton told a story about Swainson's Thrush and Raven:

> Swainson's Thrush invited Raven to her house for a meal. She told her kids to take their baskets out to pick berries. She started singing her song. In her song, she nicknames each of four colours of salmonberries and then sings their common names:
>
> NENELḰXELIḰ (*nənəl'q'x̱əliqw*, "the little black/dark red-headed ones"),
> NENELPKIḰ (*nənəl'pq'iqw*, "the little white-headed ones"),
> NENELḰEMEḰ (*nənəl'kwəməqw*, "the little red-headed ones"),
> NENELPW̱IḰ (*nənəl'pxwiqw*, "the little blond/golden-headed ones"),
> W̱EW̱ELEW̱ELEW̱ELEW̱ES̱! (*xwəxwələxwələxwəlxwəsh*, "ripen, ripen, ripen, ripen!").
>
> As she sang, her children's baskets filled up.
>
> Afterwards, Raven said, "You come to my house." Swainson's Thrush did. Raven told his children to go out with their baskets. They did that for their dad. Raven sang and sang in his croaking voice, but the baskets never got full.

Trailing Blackberry   *Rubus ursinus*
Berries: SKELÁLṈEW̱ (*sqw'əlélngəxw, sqw'əlngəxw*)
Vines: SKELÁLṈEW̱IⱠĆ (*sqw'əlélngəxw-ilhch*, sqw'əlngəxw-ilhch)

This native species is a small, slender cousin of the large, robust Himalayan Black-berry (*Rubus discolor*) that is now common throughout W̱SÁNEĆ territory and over much of Vancouver Island and the adjacent mainland. Trailing Blackberry is a bluish-stemmed spiny vine that grows over bushes and trails along the ground in tangled masses. Some of the stems are many metres long, and they can root at the tips or places where they touch soil. The leaves have three, four or five leaflets, each one pointed with sharp teeth around the edges and more-or-less spiny along the veins underneath. The flowers are white with five elongated petals. Male and female flowers are borne on separate vines. The berries ripen early compared with those of Himalayan Blackberry, beginning in late June or early July, although they can be harvested later at higher elevations. The berries change from green to bright red to shiny black as they ripen. They are sweet and flavourful. The leaves usually remain on the vines over the winter, but often turn a deep red colour.

Trailing Blackberry is common in woods, clearings and thickets throughout W̱SÁNEĆ territory from low to mid elevations.

Traditional Use: People still pick these berries when they can find them. They were Elsie Claxton's favourites. Dave Elliott noted that the W̱SÁNEĆ used to dry large quantities of SKELÁLṈEW̱ in cakes for winter. Chris Paul said that people used to store the dried cakes in alder-wood crates. To prepare these dried berries for eating, you cut off a piece and put it in hot water for a while. You can also add a little sugar as a sweetener. Nowadays, people more commonly pick the berries of the larger, more common Himalayan Blackberry.

Chris and Elsie said that SKELÁLṈEW̱IⱠĆ leaves, especially the red ones, were used to make tea, and can also be added to various medicines, as a sweet-ener. Blackberry leaf tea has a pleasant fruity flavour. Some people recommend it for a poor appetite. Mary Thomas said that its leaves are good for washing your eyes, if they are weak.

Both Chris and Dave remarked that, before new dancers would dance, they would scrub their bodies with the stems of this bush as a purification ritual. It is a highly respected spiritual plant.

Tim Montler recorded a story about the origin of blackberries from Elsie Claxton, Vi Williams, and Manson Pelkey (Elsie's nephew). In this story, a young woman climbed to the top of a pointed cedar snag (for various reasons, depending on who's telling the story: she wanted to look for her brothers or her mean husband made her go up or she saw something shiny that she wanted). She became impaled on the top. She cried out for help to her brothers who she could see paddling in a canoe. The younger brother in the bow said, "Can you hear our sister calling?" and the older brother in the stern said, "I don't hear anything, that's just your lice talking to you." The woman survived, but blood flowed down the snag into the bushes where it became blackberries. That's why you find blackberries growing on cedar stumps.

Ethnographer Diamond Jenness (n.d.) recorded the following story about Raven and his craving for blackberries:

Xwabic, the vireo [identity uncertain], invited her sister birds and beasts to accompany her to pick blackberries. Raven went with them to protect them. They paddled in Xwabic's canoe to the berry ground, where they filled their baskets and carried them down to the beach. There, Raven began to groan that he was sick (he had done nothing but eat berries).

He said, "I will go up the hill and gather some moss. That may help me!" He filled his blanket with moss.

They loaded the baskets into the canoe, and started out, Raven steering. As they travelled a fog came up and enveloped them. Raven took bunches of his moss and laid it on the water, where it swelled up and through the fog had the appearance at a distance of canoes.

He called out, "Paddle hard, my sisters, the enemy are pursuing us. Paddle hard."

They paddled hard and reached a cliff. "Flee up the cliff, my sisters," cried Raven. "I will stay and fight the enemy."

They fled up the cliff, all but Slug, who travels very slow and therefore took shelter under a clamshell on the beach. Then Raven ate the blackberries, and poured the juice over his head to pretend he was wounded. His "sisters" came back after a time, one by one, to find out what had happened. Raven lay groaning on the beach with a stomach ache.

They said, "What has happened?"

Raven answered, "The enemy beat me up terribly. See my bloody head."

As they commiserated over him, Slug came forward and said, "Alas. My brother has over-eaten!"

"Take her away!" cried Raven. "She makes me worse instead of better!"

Slug, however, repeated her refrain. Her sisters examined Raven more closely and found that the blood was only berry juice. They looked from one to the other, then quietly entered the canoe and paddled away, leaving Raven to groan alone on the beach....

Tim Montler noted that in the 1970s, Dr Earl Claxton Sr and Manson Pelkey recorded a version of this story from Baptiste Jimmy (reputed to be 100 years old at the time) in Saanich. This is considered a very funny story; everyone knows that Raven can be up to no good when he volunteers to help.

Red Elderberry	*Sambucus racemosa*
Blue Elderberry	*Sambucus cerulea*

Berries: red, ȾIWEK (*t'thiwəq'*)
 blue, Ⱦ⅄I₵I₵ (*t'thikwíkw*)
Bush: red, ȾIWEK IȽĆ (*t'thiwəq'-ílhch*)
 blue, Ⱦ⅄I₵I₵ IȽĆ (*t'thikwíkw-ílhch*)

Red Elderberry is a tall, bushy shrub that grows up to three metres tall. The greyish pithy stems and twigs are marked by raised spots called lenticels. The leaves are large and compound, with five to seven elliptical, pointed leaflets arranged on either side of a central axis. The leaves, bark and roots, when crushed, produce a pungent smell. Numerous creamy white flowers cluster in dense pyramid-shaped heads. Flowering usually occurs in May to June, and then, by July, generous clusters of shiny red berries appear. Blue Elderberry bears flat-topped clusters of powder-blue berries that ripen in summer.

Red Elderberry typically grows in moist open settings such as the edges of swamps or along floodplains. It can also be found in moist forests from low to mid elevation, and is relatively common in W̱SÁNEĆ territory. Blue Elderberry is more common in the Duncan area and north to Courtenay on Vancouver Island, but Elsie Claxton knew of it as well.

Red Elderberry.

Traditional Use: Although ȾIWEK are edible when cooked, the W̱SÁNEĆ people did not generally eat them, according to Elsie Claxton and Violet Williams. ȾȽIȻIȻ, when available, were more commonly eaten, after being cooked, according to Christopher Paul. Recently, people have mixed both kinds of elderberries with sugar and sometimes with other fruits to make jelly or jam. Violet also reported that people used to make wine from ȾIWEK. She also recalled having taken a medicine made by boiling ȾIWEK IȽĆ bark in water. This was to speed up childbirth when Violet was in labour with her daughter. She credits this medicine with having saved her life.

Christopher Paul said that Saanich children sometimes hollowed out the stems of ȾIWEK IȽĆ to make blow-guns, but these should not be used when the wood is still fresh (see Warning).

Warning: The roots, stems, bark, leaves and unripe, uncooked fruit of elderberries contain poisonous glycosides and other substances that may cause nausea, vomiting and diarrhea. Medicinal tea from the plant may cause poisoning. Children have been poisoned when they used the hollowed out stems for peashooters. You should never use these plants for medicine without the advice of a trained herbal specialist or doctor.

Blue Elderberry.

Soapberry
Shepherdia canadensis
Berries: SX̱ÁSEM (sx̱wésəm)
Bush: SX̱ÁSEM IȽĆ (sx̱wesəm-íłhch)
Other names: Soopollalie, Canadian Buffalo-berry.

Soapberry is a medium to tall, many-stemmed shrub that grows up to three metres tall. The roots bear nitrogen-fixing nodules, enabling Soapberry to successfully invade raw disturbed or recently burned ground. The bark is smooth and grey-brown. The leaves, usually 2.5 to 5 cm long, are oval, smooth edged and often somewhat droopy. Distinctive rust-coloured spots cover the young twigs, buds and especially the backs of the leaves. With the aid of a hand lens, you can see large star-shaped hairs attached in clumps to the surface of the leaves. In the spring, before the leaves emerge, clusters of inconspicuous greenish to dull red flowers appear, with male and female flowers on separate plants. The female

plants produce bright orange-red trans-lucent fruits, which usually ripen in the first half of summer, depending on elevation and climate. The berries, which are covered with small brown spots, are juicy but bitter. Some female plants bear only a few berries, whereas others may be dripping with fruit.

Soapberry occurs here and there in W̱SÁNEĆ territory. It grows in dry woods, and prefers limestone soils. The bushes are common on Malahat Ridge, around Bamberton and on the slopes above, and also occur on the Saanich Peninsula and East Sooke in some locations, as well as on the San Juan Islands.

Traditional Use: The W̱SÁNEĆ people, like most other First Peoples in British Columbia, make a favourite dessert from SX̱ÁSEM, often called "Indian ice-cream". The berries are picked in the summer, crushed in water and then whipped using bundles of grass, maple leaves or Salal branches to form a stiff froth. Fresh sweet berries, cooked camas bulbs and, more recently, sugar can be added to sweeten the slightly bitter mix. The whip was usually eaten with a special type of flat wooden spoon.

Elsie Claxton used to pick SX̱ÁSEM. She said that the bushes used to be very plentiful on the ridge above Tsawout Reserve, but that they had disappeared by the 1990s. She knew they also grew at the top of the Malahat, but did not know where to go and pick them in any quantity. People used to travel by canoe across Saanich Inlet to near Bamberton and hike up the hill just to collect SX̱ÁSEM. They also picked strawberries and harvested wild onions on these slopes. The south side of Mitchell Bay on San Juan Island is called SX̱ÁSEM, according to Dave Elliott.

People also collected and dried SX̱ÁSEM for winter use. Many people today obtain the berries from friends and relatives in the interior, where they are generally more plentiful.

Hardhack   *Spiraea douglasii*
TÁ,ȾEȽP (*teet'th-ə'lhp*) or possibly TÁ,ȾEȽP (*te7təth-lhp*), borrowed from Quw'utsun'
Other name: Douglas's Spirea.

Hardhack is a wiry deciduous shrub growing up to 2.5 metres high and often forming dense thickets. The bark is a light brown and the stems are thin and branching. The common name likely relates to its toughness – hardhack thickets are difficult to chop through with hand tools. The leaves are elongated and rounded or somewhat pointed at the tips, with coarse teeth around the margins on the upper end. In dry sunny sites the leaves may be only 2.5 cm long, but on vigorous young shoots they easily reach 10 cm. The flowers are small and dark rose coloured, forming, dense, showy elongated clusters at the tips of the stems from June to August. The seed capsules are small, dry and reddish-brown; they persist on the bushes over the winter.

Hardhack thrives in moist settings. Typical habitats include lake margins, stream banks, ditches, swamps, bogs and wet meadows. Poorly drained lowlands in the Fraser River valley harbour thousands of hectares of hardhack thickets.

Traditional Use: The straight, tough branches are the best type of wood to use for salmon spreaders when barbecuing or smoking salmon. Violet Williams and Mary Thomas said that their mother never used cedar for this purpose, only the sticks of TÁ,ȾEȽP.

TÁ,ȾEȽP wood also used to be hardened in a fire to make blades, halibut hooks, cambium scrapers and other tools (Barnett 1955). It is considered to be similar in strength and hardness to its relative, Oceanspray.

Elsie Claxton said that the brown fruiting tops were once used to make a tea to stop a prolonged menstrual flow. A girl or woman would drink the tea every day until the flow stopped. Elsie noted that people could also take this tea for any kind of haemorrhaging.

Common Snowberry   *Symphoricarpos albus*
Berries: PEPKIYOS (*pəpqʼəyás*)
Bush: PEPKIYOS IŁĆ (*pəpqʼəyas-ílhch*)
Other name: Waxberry.

Common Snowberry grows as a bushy deciduous shrub up to two metres tall. The twigs are smooth, thin and wiry. The leaves are oval and smooth-edged or sometimes deeply lobed, especially the leaves of the new shoots. The tiny bell-shaped flowers are pinkish to whitish, and are borne in dense clusters at the ends

of the twigs from late spring through summer. The fruits are clusters of soft, white berries that are not good to eat; they often persist on the plant well into the winter, after the leaves have dropped off.

Snowberry is a common shrub that often forms dense thickets, together with Nootka Rose, June Plum and other shrubs at the edges of meadows and ravines and on rocky outcrops, especially under Garry Oak and Douglas-fir in W̱SÁNEĆ territory. It occurs from sea level to mid elevations and from moist to dry sites.

Traditional Use: Elsie Claxton said that PEPKIYOS are poisonous. Chris Paul, too, said that they were poisonous and recalled that one of his sisters was fatally poisoned from eating them when he was a boy.

Elsie used to make salmon spreaders from the straight sticks of PEPKIYOS IŁĆ, for drying salmon. The sticks were also sometimes used for clam skewers. Violet Williams recounted a time when a child who was paralyzed in his legs was successfully treated by being bathed in a solution of this plant. The boy had to crawl everywhere, but after bathing in this solution for many months, he eventually started to walk. Elsie Claxton added that the branches could be placed in the bath of any person, especially a child, to make his legs strong.

Elsie also recalled that for a really bad itch you could scrape the bark off PEPKIYOS IŁĆ stems, put it into really hot water and bathe in it. She said that when they were youngsters, everyone used to get the itch. (Possibly this is "swimmers' itch", from swimming in some lakes in the summertime, when a small parasitic fluke from snails gets under your skin.) PEPKIYOS were also used as medicine, rubbed directly on rashes, sores and burns (Turner and Bell 1971).

According to Dr Earl Claxton Sr, PEPKIYOS translates as "little white revenge berries".

Blueberries
Canada Blueberry
Vaccinium myrtilloides
Bog Blueberry
Vaccinium uliginosum
and others
Berries: MOLSEN (*mal'səng*; meaning "blueberry")
Bush: MOLSEN IŁĆ (*mal'səng-ilhch*)

Elsie Claxton said that the type of blueberry she called MOLSEN is the same kind that are found in stores, and that are grown commercially in the Fraser River valley and elsewhere. She said that people used to pick them wild there, too. She saw some up in the Cascade Mountains when they went to Yakima when she was a girl, and said they were nice round blue berries on bushes about a metre high. She said they only grow on the mainland.

There are probably several species that would fall under this name, but the most common one would be Canada Blueberry, which was common in the

Bog Blueberry.

Fraser River valley, in the territory of the Sto:lo people. Canada Blueberry is a low deciduous bush that grows in dense patches. The leaves are oval to elliptical, smooth-edged and velvety. The flowers are greenish white or sometimes pinkish, and urn-shaped. They grow in short terminal clusters. The berries are borne in clusters, and are blue with a whitish waxy coating, and sweet tasting. Commercial blueberries are derived from this and related species.

The same name may also apply to Bog Blueberry, which grows in bogs on Vancouver Island as well as the mainland. Its bushes are short and bluish-green, and the berries are large and spherical, but borne singly rather than in clusters. The blueberries Elsie Claxton saw in the Cascades were probably Black Mountain Huckleberries (*Vaccinium membranaceum*) or perhaps Oval-leaved Blueberries (*V. ovalifolium*; see next page).

Canada Blueberry and Bog Blueberry grow in moist places, particularly acidic peat bogs, with Labrador Tea, Bog Cranberry, Lodgepole Pine and sphagnum mosses. Canada Blueberry may have been introduced to the Fraser River valley as a commercial crop. It grows in the Kootenays and in eastern Canada.

Traditional Use: Dave Elliott wrote about this berry. He said that it grows mainly in sphagnum bogs and that the berries were eaten fresh or dried. It would have been found growing with Bog Cranberry. People probably went to places like Burns Bog on the mainland to get them.

Oval-leaved Blueberry
Vaccinium ovalifolium

Berries: ŁEW,ḴIM (*lhəw'qím'*)
Bush: ŁEW,ḴIMIŁĆ (*lhəw'qim'-ílhch*)

Oval-leaved Blueberry is a deciduous shrub, growing up to 1.5 metres tall. The leaves are thin, oval and smooth-edged. The flowers appear before the leaves have expanded, and are solitary, pinkish-white and urn-shaped. The berries are of

good size and flavour, usually covered with a waxy "bloom" which makes them appear light blue. A similar species, probably called by the same name (ŁEW,ḴIM), is Alaska Blueberry (*Vaccinium alaskaense*). Its leaves are usually more pointed, and its berries are darker coloured and somewhat acidic.

Oval-leaved and Alaska blueberries are common in moist coniferous forests and along shaded stream banks along the coast of British Columbia, but mainly in the wetter regions. They are not common in W̱SÁNEĆ territory, except at higher elevations on the Malahat; they occur commonly westward, toward Sooke and Jordan River.

Traditional Use: The W̱SÁNEĆ people ate ŁEW,ḴIM whenever they could be obtained, and probably dried them in cakes for winter use whenever possible. Elsie Claxton and Violet Williams both knew about these berries and liked them.

Evergreen Huckleberry
Vaccinium ovatum

Berries: YIYXEM (*yi7xəm'*)
Bush: YIYXEM IŁĆ (*yi7xəm'-ílhch*)

Evergreen Huckleberry is a dense, leafy shrub that grows up to four metres tall. The leaves, mostly two to four centimetres long, are dark green, leathery and shiny, with small, even teeth along the margins. The small, pinkish-white, urn-shaped flowers bloom in early summer in clusters at the tips of the branches and along the nodes. The berries are small and spherical, usually less than a centimetre across. They have two colour forms: glossy black and powdery blue, each borne on different bushes, though the bushes may grow close together. This is one of the few evergreen species of huckleberries and blueberries (*Vaccinium*).

Evergreen Huckleberry grows in moist coastal coniferous forests. It needs a soil rich in humus and requires moisture even in the summer, so it prefers partially shady sites. It is never found growing far from the ocean shore. It is not common in W̱SÁNEĆ territory, but there are a few sites where it can be found. It is more common on the west coast past Sooke, and farther up the east coast of Vancouver Island, such as around Hornby Island.

Traditional Use: YIYXEM taste slightly tart or acidic, but they are very flavourful. Violet Williams' mother used to pick them for jam. She harvested them near Jordan River since they do not grow in very many places. These berries were probably mostly eaten fresh, since they keep very well even after being picked.

YIYXEM are generally the last of all the fruits to be harvested at the end of the seasonal round. Sometimes they can be picked right up until the end of December. For this reason, some people call them "winter huckleberries".

Florists sometimes seek the attractive sprays of evergreen huckleberry for flower arrangements. Another, similar shrub also used by florists is Falsebox (*Paxistima myrsinites*). It doesn't have berries, but has small evergreen leaves like YIYXEM IȽĆ. Elsie Claxton used to pick Falsebox greenery for florists, and was paid 25 cents per bunch. She did not know of a name for this shrub. She said that many people picked it, from the Gulf Islands and "everywhere". The going price, as of the early 1990s was $1.50 per bunch.

Red Huckleberry
Vaccinium parvifolium

Berries: S,ḴEḴĆES (*sqw'ə'qwchəs*)
Bush: S,ḴEḴĆES IŁĆ (*sqw'əqwchəs-ílhch*)

Red Huckleberry is a delicate green-twigged shrub that may reach four metres tall, but usually grows to one or two metres. The branches are flexible and sharply angled, mainly erect but spreading outwards at the ends. They bear thin, small oval to oblong leaves. The older plants lose their leaves in the winter, but in very young plants, the leaves remain over winter. Small pinkish urn-shaped blooms develop along the twigs from April to June. The berries ripen into bright red glistening globes, some up to a centimetre across. They are tart but juicy – a favourite food of children, bears and birds.

Red Huckleberry grows throughout W̱SÁNEĆ territory, in the raw humus of the moist floor of closed to partly open forests and along the edges of woodlands. The bushes are often seen perching on top of old rotting stumps as if surveying the interior of the woods. Several bushes may line an ancient log on the ground. The plants are widely spread by forest birds that eat the berries and distribute the seeds in their droppings.

Traditional Use: The W̱SÁNEĆ people picked large quantities of these juicy, flavourful berries. Expeditions were often made to gather S,ḴEḴĆES and other fruits on the Malahat. The berries were picked by hand, shaken off the bush onto mats, or removed from the bush using a scoop with wooden teeth. People sometimes cleaned the berries of leaves and debris by rolling them down a wet board.

Belinda Claxton holding a branch full of Red Huckleberries.

People sometimes ate the berries fresh, but they usually mashed large quantities and dried them into cakes for winter use. They sometimes boiled the berries, mixed them with red salmon spawn and covered the mixture with oil or grease and Skunk-cabbage leaves to seal them for later. A long time ago, people used to pick S,ḴEḴĆES into a large pack basket carried on the back by means of a tumpline over the forehead. Nowadays people still enjoy these berries, but most people use ice-cream buckets as their containers.

Herbaceous Plants

Yarrow
ṮELIḰ EⱢP (*tl'əlíqw-əlhp*)

Achillea millefolium

Yarrow is an aromatic perennial herb that grows up to 100 cm in height. The greyish-green fern-like leaves are finely divided. Most of the leaves are clustered at the base of the plant, and these are the biggest, around 10-15 cm long. The flowering stem bears smaller leaves arranged alternately on the stalk. The entire plant appears greyish because it is covered by small hairs. The flowering heads are white or sometimes pinkish, crowded together in flat-topped clusters. They bloom during the summer, after which the tops dry out and turn brown. Each cluster produces many one-seeded smooth, flat fruits. The aromatic quality of the plant is due to the presence of a mixture of volatile oils, including thujone and menthol.

Yarrow prefers well-drained, open sites such as roadsides, upper beaches, meadows and rocky slopes. It grows from low to high elevations and often becomes weedy at low elevations. It is common throughout W̱SÁNEĆ territory.

Traditional Use: Elsie Claxton's mother used ṮELIḰ EⱢP leaves to treat sore throats and colds. She washed the leaves, then gave them to the child or adult with the cold, to be chewed. The person didn't swallow the whole leaf, but only the juice. Mary Thomas said that you could place ṮELIḰ EⱢP root on an aching tooth to kill the pain and numb the gum tissues. She noted that the plant looks like little carrot leaves. Elsie knew of a similar white-flowered plant, Pearly Everlasting (*Anaphalis margaritacea*), that her mother had told her was also good for colds when you chewed the leaves, but Elsie could not recall the SENĆOŦEN name for it.

Vanilla-leaf
SEḰ,ŚEN (*sə'qwshən*)

Achlys triphylla

Vanilla-leaf is a leafy perennial often found in dense patches along the forest floor. It spreads by branching underground stems called rhizomes. The bright green leaves are all basal, growing singly from tall slender stalks up to 30 cm or so high. Each leaf is divided into three sections, all fan-shaped and bluntly toothed around the ends. The flowers grow clustered in white bottlebrush-like spikes arising from the rhizomes and extending above the leaves on a long stalk. The plant has a va-nilla-like fragrance when bruised or dried, due to the release of a sweet-smelling compound called coumarin.

Vanilla-leaf grows well in moist and shaded forests, in openings and at forest edges. It is usually quite common along stream banks at low to mid elevations. It is widespread in W̱SÁNEĆ territory.

Traditional Use: Elsie Claxton and Violet Williams saw this plant growing on the forest floor at Goldstream. Vi noted that it was used as a childbirth medicine. A woman in labour would drink a tea from SEḰ,ŚEN leaves to make the baby come more quickly. Christopher Paul and Elsie Claxton noted that the leaves can be hung around the house and are said to keep the flies and mosquitoes away. Insects apparently are repelled by the sweet odour the leaves emit when they are wilting or dried.

Nodding Onion   *Allium cernuum*
SḴEX̱ or ḴEX̱IEĆ (*sqw'ə 'x̱w* or *qw'əx̱wíyəch*; "pertaining to underarm odour")

Nodding Onion is easily recognized by its onion-like odour if you crush the leaves or stems, and by its grassy leaves and pink, nodding flower heads. The bulbs are long and narrow with a flat plate at the base from where the true roots grow. The bulbs divide readily, so that often you will see a cluster of plants growing together. The leaves are somewhat succulent and grass-like, growing up to about 20 cm in length. The flower stalks are often taller than the leaves, some reaching up to 40 or 50 cm. The stalks bend over at the top, so that the flower head nods. The flowers are pink and showy, in round-topped clusters of 10 or more on one head. Flowering occurs from May to August depending on the elevation. The mature papery seed capsules release hard black seeds when ripe.

Nodding Onion thrives in open sites (such as coastal bluffs), in dry open woodlands and on gravelly beaches above the tideline. It is often associated with Douglas-fir and Garry Oak. Its distribution is somewhat patchy, but it can be found in many locations in W̱SÁNEĆ territory.

Traditional Use: Christopher Paul said that W̱SÁNEĆ people used to paddle or row across Saanich Inlet to harvest SḴEX̱ from the Bamberton area before the cement works were built. The bulbs could be washed and eaten raw. But more often they were cooked and eaten with other foods due to their strong flavour.

Elsie Claxton remembered eating SḴEX̱ only once, at a small rocky island owned by the band called W̱AW̱TEȽ (*xwəxwtəlh*). She recalled that this island was near Sidney Spit and that it had only one tree on it and lots of seagulls. She

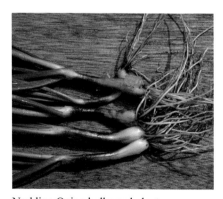

Nodding Onion bulbs and plant.

Hooker's Onion.

said there was lots of camas growing there as well.

There are other species of edible wild onions in W̱SÁNEĆ territory, recognizable by their smell. The most notable of these is Hooker's Onion (*Allium acuminatum*), whose flowers grow in upright heads and are deep purplish-pink. Elsie used to harvest these onions at ṮIX̱EN (*ts'ix̱wəng*; Cordova Spit at Tsawout) and eat the bulbs, as her daughter, Belinda Claxton, recalls. Domesticated onions are called by the same name as the wild ones.

Warning: Meadow Death-camas (*Zigadenus venenosus*) is another plant in the Lily Family that produces bulbs, but these are deadly poisonous. They have no onion odour, and their creamy white flowers grow in elongated clusters (see the photograph on page 124). Anyone wishing to sample bulb-bearing wild root vegetables must be positive of the identity of the plants they harvest.

Blue Camas
Great Camas — *Camassia leichtlinii*
Common Camas — *Camassia quamash*
ḴȽO,EL (*qwlhá7əl*) or SPÁNW̱ (*spéenxw*)

These two species are very similar, but Great Camas is taller than Common Camas, and it blooms about two weeks later. Great Camas can grow up to 70 cm tall; Common Camas usually grows to about 35 cm. These plants are in the Lily Family, and have edible bulbs located deep in the soil. Bright green leaves grow from the base of the plant and appear grass-like; Great Camas leaves grow up to 50 cm long. The flowers of both species are usually deep blue, but occasionally pale blue or even white. Plants usually bear many flowers, all six-petalled and arranged in an elongated spike at the top of the flowering stalk. Individual flowers of Great Camas are perfectly star-shaped and up to 3.5 cm across. Common Camas flowers are somewhat asymmetrical, with one petal pointing downwards and the other five pointing outwards and upwards. The seed capsules are elongated and three-sided in Great Camas, and barrel shaped in Common Camas. The capsules split into three sections to release black, glossy seeds. After the seeds have germinated, it takes several years for the plants to grow old enough to flower.

The blue camas species can be found on grassy slopes and moist meadows, at low to mid elevations. Common Camas is found throughout southeastern Vancouver Island north to Comox and west to Port Alberni; Great Camas grows mostly in the Victoria area.

The two species of blue camas: Great (left) and Common.

Traditional Use: ḰŁO,EL was the most important root vegetable for the W̱SÁNEĆ people. It was the only widely available source of carbohydrate in a diet that consisted mainly of fish and meat. The W̱SÁNEĆ used to dig much of their ḰŁO,EL bulbs from many of the smaller Gulf Islands, such as Mandarte and Arbutus islands. Areas over rock, such as along the rocky cliffs by the sea, were preferred harvesting sites because the bulbs were not too deep. ḰŁO,EL bulb beds were generally divided into family plots, which were passed down through the generations. These plots were kept clear of stones weed and brush. The bulbs were usually dug between ĆENŦEḴI and ĆENŦÁ,WEN (June and August). The entire family, including men, would be involved. The harvest usually lasted for several days. Sometimes harvesting was done in connection with fishing trips to the islands or to Boundary Bay, as recalled by Dave Elliott. The bulbs were dug up with a pointed stick of Yew or Oceanspray and placed in baskets carried on the back with a tumpline over the forehead. The soil was lifted out in small sections, and only the largest bulbs, at least five centimetres across, were removed, according to Dave Elliott. He and Elsie Claxton recalled that the bulbs were collected when the seed pods were dry. This is when the bulbs are the largest and most nutritious. Dave said that the large flowering beds were marked so the bulbs could be located later. Broken stems with the mature seedpods of ḰŁO,EL were sometimes buried in the loosened soil as people were harvesting the bulbs. Bulb beds were usually burned every year after the harvest to increase yield in the following year.

Marguerite Babcock describes camas harvesting in an unpublished article that she wrote in 1967:

As of 150 years ago, camas was still the "number one" vegetable of the Indians; Christopher Paul's mother's mother also told him that in her youth, camas was still being gathered with great frequency. [His mother was Halkomelem.]… [Christopher Paul] accounts for the decline in emphasis on camas gathering in the post-white years by the introduction through the whites of growing potatoes, carrots and other vegetables, which began to demand more of the Indians' time than simply feed-gathering activities did. However, the newly developed agriculture on the Saanich Peninsula cut down on the number of camas plants available. By the time of Christopher Paul's youth, camas was no longer gotten "wholesale", and was not an everyday food, being served as far as he knew mostly at special occasions such as the big winter invitational dances. Chris Paul is sure that East and North Saanich and Kuper Island [Penelekut] Indians were getting camas in his youth, but that the West Saanich Indians were not; he is not certain though, of what other Indians may or may not have been gathering it at the time. No Indians gather it today, he stated.

Christopher Paul knows through his mother's mother that at one time the Indians who gathered camas would establish plots for this purpose, each family claiming a plot or plots. Before the advent of white settlers, the Indians established these plots in the general areas where their own groups lived, Christopher Paul thinks. And for several decades after arrival of the whites, they could continue to gather it from both non-reserve and reserve land, he supposes, since for some time (even up to the time of his youth) the white settlements were still few and scattered. All the while, the Indians were also gathering camas from the islands in the Strait of Georgia (James, Salt Spring, Sidney, Saturna, etc., islands), although he doesn't know if the families also established plots on these islands…. By the time of his youth, the range of camas gathering may not have diminished much since the time of his mother's mother's younger years, but the frequency of the gathering was much lower. Another change had also occurred: as far as he could tell, there was no more establishing of plots to gather camas by the time he was a boy. He made this remark on the basis of the fact that never did he actually see any cleared plots, although he knew that camas was still being gathered in his lifetime since he himself had eaten some….

When the plots were still being laid out for purposes of gathering camas, each family respected other families' plots. Moreover, reserves, once they became instituted, respected each other's boundaries in regard to camas gathering. However, if an Indian from one reserve was closely related to Indians living on another reserve, he might freely gather camas on the land of that other reserve….

When each family claimed a plot of land … it was concerned only with the camas itself, and not with the actual land. The family would keep a claim on its plot or plots as long as the earth there held useable bulbs, which might be for only one picking; however if the family left some small, immature bulbs in the plot which could be picked the next year … the family might have kept claim to the plot for more than one picking season.

The way that the family group ... would establish claim to a plot of land [for camas harvesting] would be by clearing it. Once a family cleared a plot, it would "just naturally" become their plot to use, explained Christopher Paul. This clearing was done in the fall or spring before the gathering season, Christopher Paul thinks; in those seasons the soil was soft from the heavy rains, but not muddy (or frozen) as in the winter.... The plot from which the bulbs were to be gathered would be cleared of stones, weeds and brush, but not of trees. The stones would be piled up in a portion of the plot where there were no

Great Camas bulbs.

camas plants growing, and the brush would be piled to one side, left to rot or to be burned.... This brush was actually uprooted, not just cut down.... The purpose of the clearing, said Christopher Paul, was to make the camas easy to clear [sic: dig?] when the camas was gathered intensively. The piles of stones on the plots are the remains or "markers" of the plots ... however, he doesn't know how or if the Indians set about marking off their plots other than clearing them.... He thinks that these plots may have been cleared every year before their use, in order to facilitate the gathering of the camas bulbs....

Christopher Paul said that [gathering camas] was done in approximately late June, when the bulbs were dry and dormant.... When gathering, the Indians would not collect the immature bulbs; Christopher Paul's mother's mother told him that these small, soft bulbs were "not worth cooking", and he has overheard his own parents discussing the matter. He thinks that these bulbs may have been gathered in another year, when they were mature, by the same family.... The mature bulbs sometimes get as big as about 2½ inches [6 cm] in diameter. After the bulbs become too old, however, they aren't any good to eat, either, Christopher Paul's maternal grandmother said.

Christopher Paul also noted that camas bulbs were often used as gifts for relatives and friends, including those up on the west coast of the island, and were distributed at potlatches, usually three or four bulbs per person.

Often, most of the bulbs collected by a family were used in a feast or potlatch ceremony upon returning to the village. The bulbs were usually cooked in enormous circular steaming pits on the beach, often 50 kg or more at once. The people dug a hole half a metre deep and just over a metre across, and placed fine dry wood in the bottom. They laid heavier sticks on top in parallel position. They placed large flat rocks on top of the sticks, then lit a fire. When the rocks became red hot, the people removed the ashes and levelled the bottom. Then on top of the rocks they put Bull Kelp blades, Trailing Blackberry vines, and Salal branches, fern fronds or Grand Fir boughs, and poured the ḰȽO,EL bulbs on top.

Sometimes they added the bark of Red Alder or Arbutus to give the bulbs a red-dish colour. They laid more kelp blades and shrub branches on top, leaving a hole so that water could be added to generate steam. Finally, they placed grass or old Cattail mats over the hole and covered the pit with about 10 cm of soil or sand, until no steam could escape. Sometimes they lit a fire on top of the pit, as well. The W̱SÁNEĆ people cooked clams, deer and porpoise in a similar manner.

After a day and a half, the camas bulbs could be removed. When cooked, ḰȽO,EL bulbs are soft, rich brown and sweet tasting. They were eaten immediately or stored in Cattail bags after drying slightly. It was said that they did not keep very long before going bad though. Fully dried, ḰȽO,EL bulbs are blackish. People traded them widely. Elsie Claxton recalled that the Nuu-chah-nulth people (of Vancouver Island's west coast) really liked camas bulbs. Long ago, they would buy them, already cooked and dried, from the W̱SÁNEĆ people for $5.00 or $10.00 per 50-pound [23-kg] burlap sackful.

ḰȽO,EL bulbs were often used to sweeten other foods. A favourite dish of the W̱SÁNEĆ people was camas whipped up with soapberries. They often ate the bulbs with salmon. ḰȽO,EL bulbs were a favourite food of Elsie's mother, but Elsie herself had never dug them up or seen them pit-cooked. The only wild roots that Violet Williams ate as a girl were the bulbs of ḰȽO,EL, which she liked very much.

Violet's mother-in-law told her this story:

There were three girls – two sisters and their cousin or friend – who were out camping with their families, digging the bulbs of ḰȽO,EL. That night, the girls were lying awake. They were looking up at the stars and talking to each other.

One girl said, "I wish I had that Bright Eyes up there for my husband."

"I wish I were married to that Red Eyes," replied another.

They soon fell asleep. The next morning, the two sisters woke up in a different country. It was very strange and beautiful.

"Where are we?" they asked each other.

Then, two young men appeared. "Who are you?" the girls asked them.

"We're your husbands," was the answer.

"But we are not even married!" they said.

"Last night, didn't you wish for stars as your husbands? We are the ones you wished for. We are star men."

The girls learned that they were up in the sky country. They were told by the star men not to dig too deeply when they went out to dig ḰȽO,EL bulbs. If a bulb broke, they were not supposed to dig down to get it. One girl was curious and she did dig deep, and they saw a little hole. They had discovered why the star men told them not to dig that deep. Through the hole, they saw another world far down below. They realized that they were looking down at their own world. They became homesick and wanted to return home. They thought about how they could get back to their own land, and decided to try to make a rope.

From then on, instead of digging ḰȽO,EL bulbs every day like they were supposed to, they started gathering SLEWI [slə'wi7] (the inner bark of cedar). They made it into xe´yexwten (cedar bark rope). They rolled the fibres between their hands and their legs to make string, and then wound the string together to make rope. They did this day after day, until eventually they figured the rope was long enough to reach the earth down below. They lowered it down through the small hole they had made, and when they brought it up, moss was on the end of the rope. This is how they knew that the rope was long enough and had touched earth.

One of the girls volunteered to go down first. She was lowered down and down. Her sister watched her going further and further; she looked like a little spider on the end of its web. Then the one sister reached the bottom, and pulled the rope as a signal to the other. The other sister followed, and they pulled the rope down after them. The rope is still lying in the mountains somewhere. [Some people say that Knocken Hill off Burnside Road in Saanich is the site where the rope came down.] From this day on, the girls never wished for star men with bright eyes or red eyes again. Now all girls are carefully looked after when they go out, and are told not to wish for stars.

There are many different versions of this story. Here is another one, recorded by Diamond Jenness (n.d.: 141) from a W̱SÁNEĆ storyteller around the 1930s:

The Sisters Who Married Stars

Two girls lay outside and watched the stars. The stars came down and carried them up into the sky. A bright-looking star took one girl; it was a handsome young man. A red watery star took the other girl; he was an old man with red eyes.

The sisters were sad. One said to the other, "Let us make a very long rope of cedar branches."

They did so. When it was finished one girl started to descend, while the other waited above. She slid down for two days, stopping at intervals to rest; her ankles were protected by some covering, but her hands were chafed almost to the bone. At last she reached the end of the rope just a little way above the earth and jumped. She alighted safely, and after resting a few minutes shook the rope as a signal to her sister.

Then the sister began to descend. She also alighted safely.

After they had rested the elder said, "Let us join another length of rope to the end of the line and swing it." They did this. The whole rope came whirling down out of the sky while they fled. Where it fell no one knows, but perhaps in some far away country.

The girls then started out to find their home. They wandered for a long time until they came to people who recognized them as the two girls who had been lost.

Warning: Be careful never to confuse the bulbs of edible blue camas with those of the Meadow Death-camas (*Zigadenus venenosus*), also known as White Camas. Meadow Death-camas bulbs are similar, but its flowers are creamy white and form a denser cluster at the top of the stem. Its seed capsules are smaller and more

Meadow Death-camas.

densely clustered as well. Chris Paul told of a sure way to tell the difference between the bulbs: blue camas bulbs are covered with a thin light-brown skin, while Death-camas bulbs are pure white. Both types of bulbs have a blackish crusty outer skin, though, and they do look very similar. True to its name, Meadow Death-camas is deadly poisonous.

Thistles

Canada Thistle	*Cirsium arvense*
Short-styled Thistle	*Cirsium brevistylum*
Bull Thistle	*Cirsium vulgare*

S,ÍYEŦ,IŁĆ (*s7əyəth-ílhch*) or XEU,XEU,EŁP (*x̱əw'x̱əw'-ílhp*)

Thistles are biennial or perennial herbs that grow from large taproots. The young, non-flowering plants produce a basal rosette of prickly leaves that spread out from the centre. The older flowering plants have leafy stems that are thick and partially succulent nearer the base. Most thistles are hairy and some grow to two metres in height. The leaves are shallowly to deeply lobed, depending on the species, and

Short-styled Thistle.

the edges are more or less spiny. Some types, like the introduced Bull Thistle, have particularly long, sharp spines on the leaves. The flowering heads of thistles are borne at the top of the plant, with whitish, pinkish or purple flowers in a dense head surrounded by spiny bracts. After the flowers are pollinated, single seeded achenes develop from each flower, each being attached to a dense cluster of fine hairs called thistledown that carries the seeds on the wind. Several species can be found in W̱SÁNEĆ territory. Two of the most common are introduced: the large-headed, dark-purple Bull Thistle, and the smaller Canada Thistle, which has more flower heads. One native species, Short-styled Thistle, has dense woolly flower heads and shallowly lobed leaves with short spines.

Thistles are common in moist meadows, clearings and forest openings throughout W̱SÁNEĆ territory. Bull Thistle and Canada Thistle are very weedy, and usually occur in disturbed ground and overgrazed pastures. Short-styled Thistle is found along roadsides and at the edges of woodlands.

Traditional Use: The large taproots of thistles were peeled and eaten raw or cooked, according to Chris Paul. Through Dave Elliott, Chris also noted: "The sharp leaves of the thistle would drive away evil spirits. They would be put in bathwater to give protection to a person before attending a feast or gathering where bad spirits might be." Sometimes, for protection, thistle roots are put in the corners of a room, or were rubbed on the face. Some people carry thistle roots in their pockets to ward off any evil.

Note: The W̱SÁNEĆ name, S,ÍYEȾ,IȽĆ is based on the word, S,ÍYEȾ (*s7əyh*), meaning "sharp". The other name, XEU,XEU,EȽP, was probably borrowed from Quw'utsun'.

Wild Strawberries

Seaside Strawberry	*Fragaria chiloensis*
Woodland Strawberry	*Fragaria vesca*
Blueleaf Strawberry	*Fragaria virginiana*

Berries: DI,LEK (*t'iləqw*)
Plants: DI,LEK IȽĆ (*t'iləqw-ílhch*)

These perennial herbs form loose patches, spreading by long, thin stolons or runners. The plants grow from short, thick rootstocks anchored to the ground by tough wiry roots. The clustered leaves are three-parted, with each leaflet coarsely and evenly toothed around the edges. The leaves of Blueleaf and Seaside strawberries are bluish green; those of Woodland Strawberry are a brighter green and more textured. Seaside Strawberry leaves are thicker and shinier than those of the other two.

In Blueleaf and Seaside strawberries, the terminal tooth of the top leaflet is usually shorter and smaller than those on either side of it. This characteristic is enough to distinguish them from Woodland Strawberry, whose

Blueleaf Strawberry.

Mature berries of Woodland (left) and Seaside strawberry plants.

terminal tooth is larger and longer than adjacent teeth. Wild strawberries bloom in April and May, the plants usually bearing several white five-petalled flowers, about 2.5 cm across, on slender stalks. As the fruit develops, the receptacle swells, presenting the seeds as little pips on its surface. The mature berries are bright-red, fragrant and delicious. Blueleaf and Seaside strawberries are usually more spherical and borne closer to the ground than Woodland Strawberries, which are elongated and borne on upright stalks often higher than the leaves.

Wild strawberries are found throughout W̱SÁNEĆ territory. Woodland and Blueleaf strawberries grow in moist meadows and along edges of roads and woods, whereas Seaside Strawberry grows in dunes and rocky crevices along the coast. Any open habitat (except bogs) supports wild strawberries. They flourish in areas that are burned periodically.

Traditional Use: Not surprisingly, wild strawberries were a favourite fruit of the W̱SÁNEĆ. DI,LEK IȽĆ patches used to be tended by burning them over and keeping them cleared of brush so that they would be more productive. DI,LEK were eaten fresh in the summertime. They were rarely dried, because they are too juicy. Christopher Paul said that an excellent tea was also made from dried strawberry leaves. The elders say that DI,LEK IȽĆ once were much more common and productive than they are today. There is a mountain on Malahat Ridge called SŦIYAS (*st'thi7yas*, meaning "like a strawberry's face") or, according to Tim Montler, just YAAS, referring to Mount Jeffery. According to Henry Smith from West Saanich, if you point at this mountain, showing it disrespect, the weather will change to rain. Tim Montler pointed at it at one time and it did start to rain. Manson Pelkey told him that if you want to single out that mountain or that direction, you should just nod your head that way if you do not want it to start raining.

Violet Williams, Elsie Claxton and their families used to go over to Orcas Island and the Mount Vernon area in Washington to pick cultivated strawberries.

Rattlesnake Plantain   *Goodyera oblongifolia*
SQEL,QELEX (*skw'ə 'lkw'ələx*; meaning "it's got spots")
Other names: Rattlesnake Orchid.

This low evergreen orchid has a basal rosette of thick, bluish-green leaves that are marked and mottled with white stripes, giving the plants a distinctive look. The leaves are oval or elliptical, about 5 cm long, pointed and smooth-edged. Rattlesnake Plantain spreads by shoots growing out from the rootstock, so that often you will see a small cluster of plants rather than just one. The dull white to greenish flowers grow in a dense spike at the end of an upright stalk that is usually 15-20 cm high. The fruits are small, dry capsules.

Rattlesnake Plantain usually grows in mossy places on moist humus in open to shaded coniferous forests. It is common, though sporadic, at low to mid elevations in many parts of W̱SÁNEĆ territory.

Traditional Use: Both Violet Williams and Elsie Claxton recognized this small forest plant when we saw it growing near Jordan River. They said, "SQEL,QELEX" ("it's got spots"), referring to the white-marked leaves. This is more of a description than a name. Elsie did not know of any particular name for it, but Vi called it *7elə7náw* in Quw'utsun. She said that the leaves are a good medicine to help clear the throat for singing or speaking.

Elsie mentioned that it was used for birth control. A tea made from the plant was taken daily to prevent a woman from having another child.

Christopher Paul said that the leaves of the plant were boiled and used as bath water for sprinters and canoe-pullers. A liniment made from the plant was said to be good for relieving stiff and sore muscles.

The plant was also used for protection. For example, Vi noted that after a funeral, a person would wash with a solution of this plant before touching a child. Sometimes it was mixed with KEXMIN (Wild Celery) in water to use as a protective wash.

Cow-parsnip   *Heracleum maximum*
YOLE, (*yálə7*)
Other names: Wild Rhubarb, *Heracleum lanatum*.

Cow-parsnip is a large, robust, leafy perennial that grows from a thick cluster of taproots. The leaves are large and hollow-stemmed. The blades are divided into three segments, each lobed and toothed. The stems are leafy, hollow and covered with rows of fine hairs. At maturity, the plants can grow up to two metres tall.

The entire plant has a pungent odour, reminiscent of celery but stronger. The flower heads are large, conspicuous umbels, with many small white flowers clustered at the tip of each spoke. Flowering occurs in late spring. The fruits, which develop in summer, are flat and about a centimetre long, with corky, thick ribs.

Cow-parsnip grows in moist open places, such as roadsides, old pastures, wet meadows, mudflats, tidal marshes and coastal shores. It grows from low to subalpine elevations and is common in W̱SÁNEĆ territory. Abundant stands occur on the Saanich Peninsula along Wallace Drive, at the site of a once far more extensive wetland.

Traditional Use: The very young stalks of both the flower buds and the leaves of YOLE, are edible. Both Elsie Claxton and Violet Williams enjoyed these shoots as girls. They said, "You just peel it and eat it like a vegetable." The stalks are juicy and have a strong celery-like taste. Sometimes they were boiled, but mostly they were simply eaten fresh and raw. YOLE, stalks had to be gathered before the flower buds opened, while they were still tightly enclosed in the leafy sheath, so

Cow-parsnip shoots.

that they resembled a covered bump on the stem. Older plants, once they started to flower, were too tough and strong tasting.

Warning: Cow-parsnip, like several other members of the Carrot Family, must be handled with care because it contains phototoxic compounds which make the skin sensitive to ultraviolet light, and therefore, to sunlight. The stalks must always be peeled before they are eaten, or they can cause blistering and discoloration of the lips and face. Light-skinned people are especially vulnerable; even brushing up against the hairs of the stem and then exposing your skin to sunlight can result in blistering and irritation.

And always be careful to ensure that you identify properly any wild plants you sample. Cow-parsnip has some very poisonous relatives, including poison hemlock (*Conium maculatum*) and water-hemlock (*Cicuta douglasii*). These plants have more finely divided, fern-like leaves, but their flowers are white and similar in many respects to those of cow-parsnip.

American Dunegrass
SŁE,QÁI, (*slhəkw'é7i*)
Other name: Dune Wildrye.

Leymus mollis

This perennial grows to 1.5 metres in height, usually in dense clumps. The tops of most stems have fine hairs. The leaves are flat to folded, 6 to 15 mm wide, and hairy on the upper side. The inflorescence forms into a spike 10 to 30 cm long, where the seeds develop to maturity.

American Dunegrass is found on coastal sand dunes and gravel beaches. It can also be found at the edge of forests near the ocean and on small islands.

Traditional Use: The W̱SÁNEĆ people wrapped the tough bluish-green leaves around the side and bottom of the leads of the reef-net, according to Dave Elliott and Elsie Claxton. Dave noted that SŁE,QÁI, twisted and shone under the water and kept fish from leaving the lead. The moving grass also camouflaged the lines of the lead, he said (see the story on page 81). Elsie recalled that people also used the leaves to tie on willow fish-traps, and that they used to take many bundles of the leaves out with them when they were reef-net fishing so they could be working on the traps while they were out in the canoes. She also mentioned that non-aboriginal people learned from aboriginal people how to use SŁE,QÁI, as a

American Dunegrass beach community.

tie. The leaf blades were also used in cupped hands to make a whistle call for attracting female elk and deer.

Another grass-like plant, called SȻELEL ("round grass"), probably Common Rush (*Juncus effusus*), was harvested in spring before blooming, according to Dave, and was used to make small fancy baskets.

Note: The general term for "grass and grass-like plants" is SAW̱EL (*sáxwəl*), according to Elsie Claxton and Violet Williams. Wheat (*Triticum aestivum*) is called LEWAN, according to Earl Claxton, who learned this word from his mother. Tim Montler pointed out that LEWAN is the Chinook Jargon word for oats, from the French *l'avoine*.

Wild Celery   *Lomatium nudicaule*
KEXMIN (*q'əx̱mín*)
Other names: Indian Celery, Indian Consumption Plant.

Wild Celery is a smooth bluish-green perennial growing from a strong taproot. At maturity, the flowering stems can reach up to 90 cm tall, but usually the plants are about 30 cm high. There may be one to several flowering stems on a plant. The leaves are basal and compound, dividing one to three times, into 3 to 30 leaflets that are oblong in shape and may have smooth or coarsely toothed edges. Tiny yellow flowers are borne in umbels, typical of plants in the Carrot Family. Flowering occurs in mid May and by July the seed-heads are starting to form. The fruits are up to 15 mm long, and are flat, with broad wings and distinctive striated ribs.

Wild Celery plants and seeds.

This plant grows throughout W̱SÁNEĆ territory in open meadows at low to mid elevations. Elsie Claxton used to collect large quantities of the seeds from Sidney Spit, where it grows in abundance. Violet Williams recalled that the plant grows beside the railroad track along Admirals Road in Esquimalt and around Langford as well, but it is not as common in these locations as it used to be. Hagan Bight near Patricia Bay is called KEXMINEN ("Place of Wild Celery"), according to Dave Elliott.

Traditional Use: KEXMIN is one of the most powerful medicines of the W̱SÁNEĆ people. The seeds are chewed and the juice swallowed for colds and sore throats; they could also be steeped in boiling water to make a tea for colds and coughs. Christopher Paul and Dave Elliott said that for headaches, people inhaled the smoke of burning seeds or sometimes the aromatic smell of crushed seeds; they might also make a poultice of the seeds and place it on the head.

The seeds also have many important ceremonial uses relating to protection of people at times of illness and death. Sometimes they are placed on the stove to burn and fumigate the house in times of sickness or death to ward off evil. The use of KEXMIN in relation to the coming of Salmon is explained in the following story, recorded by Diamond Jenness (n.d.: 94) from a W̱SÁNEĆ storyteller around the 1930s:

Origin of Salmon

Once there were no seals and the people were starving; they lived on elk and whatever other game they could kill. Two brave youths said to each other, "Let us go and see if we can find any salmon." They embarked in their canoe and headed out to sea, not caring in what direction they travelled. They journeyed for three and a half months. Then they came to a strange country. When they reached the shore a man came out and welcomed them, saying, "You have arrived."

"We have arrived," the youths answered, though they did not know where they were.

They were given food to eat, and after they had eaten their host led them outside the house and said, "Look around and see what you can see."

They looked around and saw smoke from KEXMIN that the Steelhead, Sockeye, Spring and other varieties of salmon were burning, each for itself, in their houses.

The youths stayed in the place about a month. Their hosts then said to them, "You must go home tomorrow. Everything is arranged for you. The salmon that you were looking for will muster at your home and start off on their journey. You must follow them."

So the two youths followed the salmon; for three and a half months they travelled, day and night, with the fish. Every night they took KEXMIN and burned it that the salmon might feed on its smoke and sustain themselves. Finally they reached *Ktces* [T̲ĆÁS (*tl'ches*), Discovery Island and Chatham Islands] where they burned KEXMIN all along the beach; for their hosts had said to them, "Burn KEXMIN along the beach when you reach land, to feed the salmon that travel with you. Then, if you treat the salmon well, you will always have them in abundance."

Now that they had plenty of salmon at Discovery Island they let them go to other places – to the Fraser River, Nanaimo, etc. Because their journey took them three and a half months, salmon are now absent on the coast for that period.

The Coho said to the other salmon, "You can go ahead of us, for we have not yet got what we wanted from the lakes." That is why the Coho is always the last of the salmon.

The young men now had salmon, but no good way of catching them. The leaders of the salmon, a real man and woman, taught them how to make S̲XOLE (reef nets), and how to use KEXMIN. They also told the young men how their people should dress when they caught the salmon, and that they should start to use their purse net in July, when the berries were ripe. So today, when the Indians dry their salmon, they always burn some KEXMIN on the fire (or on top of the stove); and they put a little in the fish when they cook it. Also, when they cut up the salmon, before inserting the knife they pray to the salmon, that they may always be plentiful.

Note: The W̲SÁNEĆ name, KEXMIN, apparently comes from SKEX, meaning "to put a curse on someone". According to Dr Earl Claxton Sr, this plant is used to protect someone from a curse and from evil thoughts.

Skunk-cabbage
ȾOȻI, (t'thákw'iʔ)
Other name: Swamp Lantern.

Lysichiton americanus

Skunk-cabbage is most readily identified by its showy flowers, which appear early in the spring, often before the leaves have expanded. The flowering part consists of a large, bright yellow sheath surrounding a club-like spadix, which contains numerous small yellow-green flowers tightly packed together around a central stalk. The leaves appear with or after the flowers – they are large, soft, waxy and tropical looking. They unfurl from the ground in succulent clusters, borne from fleshy whitish, branching rhizomes. In a good site, the leaves can grow to 1.5 metres long. By late spring, the yellow sheaths die back, and the flowers ripen to produce greenish to reddish fruits embedded in the spadix. The entire plant has a pungent, skunk-like odour, which is especially attractive to the small beetles that pollinate the flowers.

Skunk-cabbage grows in rich black soil in extremely wet areas such as swamps, marshes and the shallow edges of ponds and creeks. It is found in such places throughout W̱SÁNEĆ territory, from low to mid elevations.

Traditional Use: The large, waxy-looking leaves of ȾOȻI, are not edible themselves, but they can be used for various household tasks, such as providing a surface on which to lay food, or for drying berries on. Both Christopher Paul and Elsie Claxton recalled that people out camping would fold the leaves in a certain way to make a temporary dipper or drinking cup. They could also be fashioned quickly into makeshift berry-picking containers, or used to line berry baskets. They were also used, along with other types of vegetation, for lining cooking pits when people were barbecuing salmon and clams. Some people used ȾOȻI, leaves as a medicine, for wrapping burns and wounds and for other purposes.

Warning: Skunk-cabbage, like other members of the Arum Family, contains tiny, sharp crystals of calcium oxalate in its leaves and other tissues. Do not eat the leaves, because these crystals become embedded in your tongue and throat tissues and cause intense, prolonged burning and irritation.

Indian Pipe   *Monotropa uniflora*
S,ŚIWE EȚ STḴÁYE (s*shiwə7 7ə tl'stqeyə7*; meaning "wolf's urine")

This unusual plant has no green leaves; it lives on decaying matter in the soil. The white, waxy stems of Indian Pipe grow in dense clumps from a dense root-crown. The young stems, until flowering, are upright, nodding at the top so they resemble a pipe or a shepherd's crook. They are covered with small triangular scales, and are usually around 15-20 cm tall. Once the flowers have been pollinated, they turn straight upright, lengthen and start to darken. The fruiting stalks are deep

brownish black. The dry, swollen capsules split open lengthwise and release minute, dust-like seeds. The dark seed stalks persist over the winter and can be used as an indicator of where the new white plants will appear the following spring.

Indian Pipe occurs sporadically throughout W̱SÁNEĆ territory. It is more abundant some years than others. It tends to grow in deep shade under coniferous trees, but sometimes grows well under Garry Oaks.

Traditional Use: This plant was well known to Elsie Claxton and Violet Williams. Elsie said that you could boil S,ŚIWE EȚ STḴÁYE and use the solution to bathe your feet and legs in if they were sore. She said it could be used in the bath to make your legs grow strong, and that it is especially good for children. Violet and her sister, Mary Thomas, said that S,ŚIWE EȚ STḴÁYE could be rubbed on the hands or limbs of people who had suffered a stroke, to help them regain the use of those parts.

According to tradition, this plant grows wherever the wolf urinates. The Nlaka'pamux people of the Fraser Canyon area also call this plant "wolf's urine" (Turner et al. 1990).

Brittle Prickly-pear Cactus   *Opuntia fragilis*
W̱EM,QI,OȾEN (*xwəm'k'wəyáthen*)

This low, succulent plant consists of egg-shaped, jointed stems that bud and branch to form dense patches. The stem segments are covered with clusters of sharp, hooked spines, arising from masses of tiny slivery spines. The stems produce attractive light-yellow to pinkish blooms in early summer; later, these blooms become small, spiny, fleshy fruits.

This cactus is common on dry hillsides and open plains of the dry interior of British Columbia, but on the coast it is restricted to dry rocky headlands along the

coast of southeastern Vancouver Island and the Gulf Islands. In W̲SÁNEĆ territory it grows around Gordon Head and on a few islands.

Traditional Use: Near Stuart Island there is a small island called W̲EM,QI,OŦEN,
according to Dave Elliott, whose name translates as "plant that clings to your
mouth" – an apt, if painful-sounding
description for this plant. It is not
known if the W̲SÁNEĆ people used
the cactus, W̲EM,QI,OŦEN, for anything. The interior peoples used to
scorch the spines off the fleshy stems,
then cook them and eat them. Some
people described them as resembling
greengage plums.

Note: The W̲SÁNEĆ name for this cactus is the same as the name for Sweet
Cicely, the next species described.

Sweet Cicely   *Osmorhiza berteroi*
W̲EM,QI,OŦEN (*xwəm'k'wəyáthen*)
Other name: *Osmorhiza chilensis.*

Sweet Cicely is a low perennial that grows from a carrot-like taproot. The basal
leaves are compound, twice dividing into threes, giving them a fern-like appearance. The leaflets are thin, slightly hairy, pointed, and coarsely toothed. The flowering stems are slender, branched and usually 30-60 cm tall. Leaves at the base of
the plant have long stalks, whereas stem leaves are short stalked. The flowers are
small and greenish-white, borne in umbels of a few each. The fruits are black and
needle-like, with bristly hairs that latch easily onto clothing and fur.

Sweet Cicely is commonly found
in open woods, along forest edges, and
in thickets and glades at low to mid elevations. It is a common, but generally
inconspicuous plant in W̲SÁNEĆ territory. It is most noticeable in summertime when its black, needle-like fruits
penetrate shoes and clothing and have
to be removed. Of course, this helps the
plant to spread itself wherever people
and animals go.

Traditional Use: Elsie Claxton identified this plant at Goldstream, where it grows prolifically. She recognized it by its sharp fruits. She said that it is good to bathe in a solution of W̱EM,QI,OŦEN (the whole plant or the root). It keeps you warm all the time, and is also used as a love charm. If your spouse is going to leave, you bathe with this plant and it will make them stay. It is also used for spiritual protection. Elsie had never heard of the roots being eaten, but the Lil'wat people of Mount Currie eat them, calling them "dry land parsnips." Berry pickers have to make sure that the sharp seeds do not contaminate their berries. If they are swallowed, they can get stuck in your throat and cause choking.

Note: The W̱SÁNEĆ name of this plant is the same as the name for cactus (see previous page), probably because both plants have sharp pointed parts.

Wild Caraway *Perideridia gairdneri*
SÁ,ȻEḴ (se7kwəq)
Other names: Gairdner's Yampah, Wild Carrot.

For many years there has been some uncertainty about the wild root vegetable that the W̱SÁNEĆ and others call "wild carrot". The name can refer to a number of different plants in the Carrot Family in different places. One of these is Pacific Hemlock Parsley (*Conioselinum gmelinii*), which grows along the coast. It may have been eaten by the W̱SÁNEĆ as it was by coastal peoples farther north. But in W̱SÁNEĆ territory there grows another native carrot-like plant with sweet-tasting roots, and it is the plant most likely called SÁ,ȻEḴ.

This "wild carrot", better known as Wild Caraway, has compound leaves with segments arranged in a featherlike fashion. The leaf segments are variable, often thin and grass-like, but sometimes broader and toothed around the edges. The stems grow up to about 60 cm high, and the taproots are usually bifurcated (two-forked). They are fleshy and white and have a sweet carrot-like smell and taste. The flowers usually start to bloom in early to mid July, and by this time the leaves have often died back. The small, white flowers grow in umbels. The fruits are small seeds similar to carrot or parsley seeds.

Wild Caraway is found in many locations around W̱SÁNEĆ territory, usually in grassy clearings and meadows. It is hard to spot, however, because the leaves blend in well with grasses, and then they die back by early summer, leaving only an inconspicuous flower head at the top of a spindly stalk. Places where it is known to grow include the grassy slopes across Saanich Inlet in the Bamberton area, the west face of Bear Hill, and the rocky areas above Brentwood Bay at Tsartlip.

Traditional Use: According to Christopher Paul and Dave Elliott, W̱SÁNEĆ people ate SÁ,ȻEḴ roots in large quantities, raw or steamed in pits. In summer

Wild Caraway.

Poison Hemlock.

they marked the places where this plant grew and dug up the roots next spring, before the leaves came up. One place was at Bamberton in times before the cement plant was built.

Note: Queen Anne's Lace (*Daucus carota*; also called "wild carrot") is the wild ancestor of the cultivated carrot. It has hairs on its stems and leaves, and the roots of the young plants are edible. The W̱SÁNEĆ call it ŚEW̱ḴÁN (*shəwqéen*), which pertains to the finely divided leaves. The same word also applies to the leaves of the introduced Yellow-flowered Common Tansy (*Tanacetum vulgare*). Both plants grow along roadsides and at the edges of meadows, ditches and disturbed sites around the Saanich Peninsula and Gulf Islands. Tansy has a strong aromatic scent, and is considered to be a good luck plant.

Warning: The Carrot Family contains many edible species – such as cultivated carrots, celery, parsnip, caraway, fennel and parsley – but also some toxic plants. One of them is Poison Hemlock (*Conium maculatum*), a common weed around Victoria and found in many parts of W̱SÁNEĆ territory. At maturity it is a tall plant with red-spotted, hollow stems and finely divided leaves, like parsley leaves. It has many white umbels and an unpleasant odour. The stems and leaves are smooth and hairless. Poison Hemlock resembles Queen Anne's Lace and it is easy to confuse them. For this reason, never try eating any wild member of the Carrot Family (or any other plant, for that matter) without first confirming its identity with someone who knows it.

Common Plantain *Plantago major*
SŁÁWEN EṮ SXEÁNEW̱ (*slhéwən 7ə tsə sx̱ə7énəxw*; meaning "mat/mattress/bed of the frog")
Other name: Broad-leaved Plantain.

Common Plantain is a low, leafy, somewhat weedy plant growing from a cluster of fibrous roots. The smooth bright-green leaves form a basal rosette, spreading out from the roots. They are oval, smooth-edged and textured with distinctive prominent parallel veins running from the base to the tip. Small green flowers crowd together in a slender pointed spike. Each flower produces a small, brownish seed capsule. The leaves die back in the winter.

This plant grows in moist meadows, around the edges of swamps, and in lawns and disturbed sites. It is common and weedy, and has a long association with human settlements. It grows throughout W̱SÁNEĆ territory.

Traditional Use: SŁÁWEN EṮ SXEÁNEW̱ is universally known as a good remedy for burns, wounds, stings, and skin infections. Elsie Claxton and Violet Williams knew it and had used it for their own families. Violet recalled one time when she used it to treat a serious burn on the leg of one of her family members. His thigh was burned so badly he could hardly walk. She gathered "frog's bed" and he chewed up the leaves and put them as a poultice on his leg, mixed with a bit of lard then covered the whole thing with one big leaf. Violet taped the leaf in place and wrapped it with a cotton cloth so that the poultice would stay in place. She changed this dressing every day, and the burn healed up within about a week.

SŁÁWEN EṮ SXEÁNEW̱ leaves are also good for treating boils. Violet said, "You know, [they] would be good for a lot of things." She once made a solution from the plant and used it as a wash for swimmers' itch. She thought it was so effective as a medicine that it would also make an excellent healing tea. She herself drank a solution of the boiled leaves for a persistent cough, and also used it to treat stomach ulcers for a family member.

Mary Thomas told Violet about a woman she knew who was going blind; her eyes were not good. She took this plant, roots and all, and every night she mashed it up and soaked a cloth in the juice. The placed the damp cloth over her eyes, tying it on, and slept with it like that. Pretty soon, a pus-like fluid came out. Then she became completely blind, but she kept on using the poultice. She figured that since she had been using that medicine, she should continue to use it, so her kids kept getting it for her. Then one day, something "just like a silk cloth" came out of her eyes. After that, she could see fine, and continued to see well until she died.

(As an experiment, we wrapped Richard Hebda's finger with a bruised plantain leaf to see if it would help draw out a thorn that would not come out, and put a bandage around it to hold it on. The thorn came out the next day.)

Raphia (palm fibre)   *Raphia farinifera*
SḰO,ȾEN (*sqwáthən*)

Raphia comes from the inner bark of a type of palm native to Madagascar.

This plant does not grow in the wild in W̱SÁNEĆ territory. The long tough strands are imported in bundles from the tropics, already processed and dried.

Traditional Use: The Raphia palm produces long flat strips of fibre. W̱SÁNEĆ women in the past bought SḰO,ȾEN in Chinatown or at florist shops in bundles for use in basketry. When Violet Williams was in school, they wove trays, using split Cattail leaves arranged in a circular pattern as a foundation, and Raphia strands as the weft. Cedar-bark strips were used for these trays in the days before SḰO,ȾEN was available. Elsie Claxton also used SḰO,ȾEN for making baskets when she was a child. To dye the Raphia they used Red Alder bark for red, onion skins for yellow and dandelion leaves for green. They also used the bark of a kind of needled tree, probably Western Hemlock, to make a dark-coloured dye. They boiled the dyeing material and soaked Raphia in the solution. Raphia was often used in dancers' costumes. Today, people still use strands of Raphia in their weaving, for strengthening the baskets and for decorative purposes – the strands come already dyed with bright colours such as red and purple.

Basket made with coloured Raphia and cherry-bark designs (RBCM 20244).

Docks
Western Dock
and others
DEMOSE, (*t'əmásə*)

Rumex aquaticus ssp. *fenestratus*

Other names: Wild Rhubarb, Coffee Plant (both used by the W̱SÁNEĆ people); *Rumex occidentalis*.

Docks are leafy herbaceous perennials. There are several species, some native to W̱SÁNEĆ territory and some introduced, all apparently called by the same W̱SÁNEĆ name. The plants grow from a deep-seated yellow taproot. The leaves

emerge in a cluster from the root-crown, and are generally long-stalked with variously oval, heart-shaped or lance-shaped blades that are dark green and sometimes wavy-edged. Western Dock has broad, oval leaf blades and succulent, often reddish leafstalks. The flowers of all docks are small and greenish, borne in dense elongated clusters on a branching head. They turn brown in fruiting, and resemble coffee grounds on a stem.

Docks are common in clearings, fields and gardens, and along roadsides and beaches. Western Dock often grows in damp meadows near the ocean.

Traditional Use: The name, DEMOSE, has also been reported to mean Cow-parsnip, but this may simply be the result of a confusion of the English name, Wild Rhubarb, which has been used for both plants. Christopher Paul and Dave Elliott said that the young stems of DEMOSE, can be cooked and eaten like rhubarb is today; DEMOSE, is also the name for cultivated rhubarb. Other First Peoples along the coast, especially the Haida, ate dock and called it by the same name as rhubarb. At first, Violet Williams and Elsie Claxton did not recognize a non-fruiting specimen of Western Dock. Later, however, Vi recalled that her grandmother's sister told her that it had once been a medicine. She said that to hurry the onset of labour and childbirth, you should dig the root of the plant "with tops like coffee" (two species of dock, *Rumex obtusifolius* and *R. crispus*, were growing in Elsie's yard at the time), wash it, cut it up, steep it in boiling water and drink the tea. Elsie added that this tea was also good to drink as a postpartum medicine. She said, "I guess it helps the pain and heals your insides". At first, Elsie and Vi did not recall a name for dock, but later, Elsie recognized the name DEMOSE right away as a name for rhubarb, and thought it must be the name applied for dock as well. When shown a flowering specimen of Sourgrass (*Rumex acetosella*), Vi and Elsie said it was "like the coffee plant, but different".

Warning: Dock roots have high concentrations of tannins, and the leaves contain oxalic acid. The leaves and leafstalks are edible, but should be used in moderation. Never eat the leaves of true rhubarb, as these are poisonous. Dock roots should not be used medicinally without advice from a qualified herbal practitioner or physician.

Wapato — *Sagittaria latifolia*
SḴÁUISELŁ (*sḵawísəl'lh*; Suttles 1951) or SḴÁUⱮ (*sqéwth*)
Other names: Arrowleaf, Wild Swamp Potato.

Wapato is an aquatic perennial that grows from a thick cluster of rhizomatous roots, which produce small tubers about the size of chestnuts. The leaves, shaped like arrow heads, grow on upright stalks in a cluster from the root crown. In deep water, the leaves are sometimes strap-shaped. The flowers are white and three-petalled, and the fruits are rounded capsules that produce corky seeds.

Wapato grows in standing fresh water and in the muddy edges of lakes and ponds. It is common in the Fraser River valley of the mainland, but rare on Vancouver Island.

Traditional Use: The W̱SÁNEĆ and Quw'utsun' people on Vancouver Island may have transplanted Wapato into their territories. But most knew it from their visits to mainland relatives, where they traded for it. The egg-shaped tubers were once cooked in underground pits. Christopher Paul knew about its use, but only a few people in 2001 had heard of it.

Bulbs of Wapato (right) and Camas.

Yerba Buena   *Satureja douglasii*
TI,IŁĆ (*tih-ílhch*; meaning "tea-plant")
Other name: *Clinopodium douglasii.*

Yerba Buena is a low, creeping vine of the Mint Family that has a distinctive, aromatic scent if crushed or walked on. The oval leaves, mostly around two centimetres long with slightly saw-toothed edges, are borne in opposite pairs along the slender, squared stems. The small, tubular flowers are whitish to pale blue with a protruding lip. These inconspicuous flowers grow in the leaf axils in early summer. The fruits are small brown capsules.

Yerba Buena trails along the mossy forest floor in open Douglas-fir or Garry Oak woodlands. It is found in suitable habitats throughout W̱SÁNEĆ territory and is especially common in shaded mossy areas of the Gowlland Range.

Traditional Use: Christopher Paul said that people made tea from the leaves; this was thought to be good for the blood (Turner and Bell 1971). Elsie Claxton recognized a sample of this plant, and knew that it was used for tea, but she knew of no SENĆOŦEN name for it, only "TI". Her mother used to call it that, but it did have a "real" name as well. Violet Williams commented that one does not see this plant much anymore. It is a small "real good-smelling" vine with quite small, round leaves, growing "in the bush". It is said to grow at Coles Bay. This plant was also used as a poultice for any wounds and scratches.

Note: Yerba Buena is a close relative of Summer Savory and Winter Savory (*Satureja* spp.). It grows well as a garden plant, and can even grow indoors as a potted plant. It roots easily from the nodes and the tips of the shoots. The first part of W̱SÁNEĆ name, "TI", was borrowed from the English "tea".

Tule   *Schoenoplectus acutus*
SȻELEL (*skwaləl'*)
Other names: Roundstem Bulrush, Hardstem Bulrush; *Scirpus acutus.*

Tule grows in dense patches from fleshy, branching rhizomes, usually in standing water or thick wet mud. The stems are cylindrical, stiff and upright, growing
to two or more metres tall and as thick as two centimetres at the base. The stem
bases are white and succulent. The upper stems are dark green and smooth outside
and pithy inside. The leaves are reduced to a few brownish sheaths at the base of
the stem. The flowers are borne in tight spikelets clustered at the top of the stems,
ripening to brown, single-seeded fruits or achenes.

   Tule grows in marshes, ditches and along muddy shorelines, such as around
Elk Lake. It occurs in fresh and brackish water at low elevations throughout
W̱SÁNEĆ territory.

Traditional Use: The W̱SÁNEĆ people harvested the long green stems of
SȻELEL, usually in July or August. They laid the stems on the ground in the sun
to dry, then used them for making baskets and mats of all sizes. Weavers soaked
the stems before using them. For Tule baskets, they coiled up the stems and sewed
them together with SḰO,ŦEN (Raphia) fibre, using a strand about a metre long.
Elsie Claxton recalled that she used to go with her mother to gather Tule stems.
Her mother used them to make baskets. She also made a very large kind of mat
called SŁÁWEN, which was hung around the walls of long houses where people
sat for dance ceremonies. These mats were warm and soft. For making the large
Tule mats, the weaver pushed a long, sharp needle, usually of Oceanspray wood,
right through the Tule stems lined up side by side, with the tops and bottom ends
alternating. Elsie Claxton recalled that her mother used white cotton strings for

The edge of a mat (right) made from Tule
stems (above).

sewing the stems together for mats. This string was the only material she knew of for sewing Tule mats, but she commented that some other type of string must have been used long ago, since cotton strings are relatively recent. Her mother used to wind up balls of this string from the hop fields at Yakima after they had picked all the hops from the vines. Few people today know how to make the big mats of Cattail leaves and Tule stems. Violet Williams commented sadly that everyone is too busy to make mats and baskets these days.

Cattail — *Typha latifolia*
SȾA,ḴEN (*st'thé7qən*; meaning "something with hair on the top")
Other name: Common Bulrush.

Cattail is a tall herbaceous perennial of marshy areas, growing from thick, whitish, branching rootstocks. These rootstocks, technically called rhizomes, send out masses of roots from the lower surface into the slimy mud of marsh environments. A stout bundle of narrow leaves shoots upward from the growing tip of

each rhizome. The smooth, light-green leaves may reach three metres tall. Inside, they are very pithy, an adaptation to growing in aquatic environments. A stiff, cylindrical flowering stalk rises from the centre of the leaves. The flower spike has two parts: the lower part bears greenish, densely packed female flowers and the upper part pollen-bearing male flowers. In the early summer these spikes release masses of yellow pollen to be carried away by the wind. After pollination takes place, the pollen flowers dry up and fall away and the lower part of the spike thickens and turns deep brown – the cat's tail. In the late fall and winter, this ruptures into a fluffy mass of downy seed.

Cattail is found in dense patches in marshes, swamps, lake edges and ditches. It flourishes in still or gently flowing water at low to mid elevations. There is a large patch of Cattail at the Goldstream Reserve and there used to be extensive patches in the lagoon at Island View Beach.

Traditional Use: The W̱SÁNEĆ people used the female (non-flowering) plants to make mats; according to Tim Montler, they did not use the male plants (those with brown cat-tail spikes). In late summer they gathered the long strap-like leaves of the female plants in enormous bundles, then separated and dried them in the sun for several days. Like Tule stems, Cattail leaves were used to make large mats, about 1.5 metres wide and 2 metres or more long.

The mats were constructed by laying the dried leaves with the tips alternating at either end to keep the mat even. The leaves were then threaded together through their sides at about 10 cm intervals with twine made of the lower edges of SṮA,ḴEN leaves or Stinging Nettle fibre. A long needle, triangular and somewhat flat, was used to thread the fibres through. It was usually made of Oceanspray wood. The needle was poked through the entire row of leaves. Once the needle was in place, a maple-wood creaser, grooved to fit the needle, was pressed around the needle, to crimp the leaves and made a channel for the thread. The needle was threaded through the eye at its tip, and then pulled back though the leaves. This process was repeated until the whole mat was sewn together. Pieces of braided SṮA,ḴEN were folded over the ends of the mat and bound to protect its edges. Tule mats were made in a similar way.

The W̱SÁNEĆ people were famous for the quality of their mats. The mats were hung on the walls for insulation in the winter, and placed on the ground in the big houses. They were also used for the walls and roofs of temporary camping houses. Both Elsie Claxton and Violet Williams had childhood memories of their mothers weaving these mats. The mats were also used as mattresses for sleeping on.

Dave Elliott recalled that the W̱SÁNEĆ people made three kinds of mats:

SȽÁWEN (*slhewən*) – a mat that covered the platforms in the house used under a mattress, most likely deer hide stuffed with feathers or piled fur hides;

KEL,ŚTEN (*q'əl'shtən*) – a shelter mat used in travel on the bottom of a canoe and to create a shelter or to sleep on; and

SOLE,EĆ (*saləʔəch*) – a mat used as a house lining to cover the walls and to create separations in the larger areas. The W̱SÁNEĆ people, like other Island Salish, probably traded SṮA,ḴEN and Tule mats to the Nuu-Chah-Nulth and the Kwakwaka'wakw peoples to the west and north. People also made woven bags from SṮA,ḴEN leaves, by spinning the split leaves on the bare thigh before twining them. Camas bulbs and crabapples were commonly stored in these bags according to Suttles (1951). When Violet Williams was in school they used to weave trays with split SṮA,ḴEN leaves as a foundation and Raphia as the weft. The warp, or foundation, was a circular pattern. The leaves were split and spun on the bare thigh before weaving. Homer Barnett (1955) reported that the woolly fluff from SṮA,ḴEN seed heads was spun with dog fur or Mountain Goat wool to make blankets. The charcoal from burning the plant was also used for tattooing.

Stinging Nettle
Urtica dioica
ȾEX,ȾEX or ȾEX,TEN (*t'thə 'x̱-t'thəx̱* or *t'thə 'x̱ən*; meaning "poison, stinging")

Stinging Nettle is a herbaceous perennial. It grows with upright stalks to more than two metres from branching rhizomes. It often forms extensive patches. Nettle stems and leaves are covered with fine, stinging hairs. The leaves grow in opposite pairs along the stem. They are triangular, or heart-shaped, on stiff stalks. The edges of the leaves are coarsely saw-toothed. The small greenish flowers grow in drooping clusters at the upper leaf nodes. The seeds, which ripen in midsummer, develop from the female flowers, which form on separate clusters above the male flowers.

Stinging Nettle grows in wet meadows, pastures and thickets, in open forests and along stream banks, avalanche tracks and roadsides from sea-level to sub-alpine elevations. It grows well in moist rich soil. Locally, the lagoon at Tsawout is named ȾEXTAĆ (*ts'əx̱tech*), meaning "bite of Stinging Nettle", because of the patches of stinging nettle in the vicinity.

Traditional Use: W̱SÁNEĆ people boiled and ate the young stems and leaves of ȾEX,ȾEX like spinach, but this may be a recent use, learned from European settlers (Turner and Bell 1971).

More important was the use of the fibrous stems for making twine, fishing lines, fish nets, duck nets and deer nets. The stems were cut in PEKELÁNEW̱ (October), split lengthwise with a bone needle, and dried for five or six days outside, then dried further over a fire. When dry, the stems were peeled and the fibres combed out. Then they were spun on the bare thigh, or with a spindle made from Bigleaf Maple wood. The threads were twisted into strong two- and four-ply twine, which was used for binding, tying and net construction. Christopher Paul said that fish nets were often dyed with the bark of Red Alder to make them invisible to fish.

Dave Elliott noted that the leaves of ȾEX,ȾEX were rubbed on the skin as a counter-irritant for bruises, aches and rheumatism. This is a painful treatment, but after the initial pain, the original aching is alleviated.

According to Violet Williams, the roots of ȾEX,ȾEX were also used as an ingredient in medicine for sore throats. Sometimes they were combined with Scouring Rush and maple leaves to make a tea; the maple leaves had to have fallen from the tree and been caught in the bushes, and not touched the ground.

Long ago, girls at puberty were brushed with stinging nettle all over their bodies so that, when they married, their husbands would never leave them.

Note: According to Tim Montler, ȾEX,ȾEX or ȾEX,TEN, meaning "poison" or "stinging", have the same root word.

False Hellebore   *Veratrum viride*
ȻE,NÁȽP (*kwən'eylhp*)
Other names: Indian Hellebore, False Green Hellebore.

False Hellebore is a perennial with upright, unbranched stems. It grows as tall as two metres from a thick rootstock. The leaves are large and elliptical, up to 35 cm long and pointed at the tip. The leaves have distinctive longitudinal creases or folds from base to tip, and clasp the stem at the base. The flowers are small, yellowish-green and star-shaped, borne in large, branching terminal clusters. The fruits are oval capsules with broad-winged brown seeds.

 False Hellebore grows in wet meadows, bogs and swamps. It is also found in moist open forests and alpine meadows. It grows most abundantly in subalpine areas. It is not common in W̱SÁNEĆ lands, but a small population grows in Goldstream, and it also occurs in upland wetlands of Malahat Ridge.

Traditional Use: ȻE,NÁȽP is extremely poisonous and has on occasion caused death when it has been used carelessly or without proper knowledge. W̱SÁNEĆ people did not usually take it internally at all. Rather, they used it to treat muscular aches and pains, by steeping the thick rootstocks in a bath and then soaking in the solution. The plant is said to be very powerful and should never be used without full knowledge and understanding of its effects.

False Hellebore leaves and flowers (right).

Warning: This plant contains poisonous alkaloids and should never be eaten or allowed near the eyes or used with open sores or cuts. Use only with full knowledge of its properties.

Giant Vetch	*Vicia nigricans* subsp. *gigantea*

Giant Vetch   *Vicia nigricans* subsp. *gigantea*
SNENW̱EŁ IȽĆ (*snən'xwlh-ílhch*) and other names – see below.
Other names: *Vicia gigantea.*

Giant Vetch is a sprawling perennial with succulent, ridged green stems and finely divided compound leaves each with 19-29 elongated leaflets. The plants can grow up to two metres tall. The leaves terminate in well-developed tendrils that cling around vegetation and help support the stems. The pea-like flowers are pink to purplish and clustered. The fruiting pods resemble small pea pods. They are green at first, but turn black as they mature. Each contains several round pea-like seeds. A similar pea-like plant, Beach Pea (*Lathyrus japonicus*), often grows with Giant Vetch. Beach Pea has larger, fewer leaflets in its compound leaves, and its flowers are larger, resembling small purple sweet peas.

Giant Vetch is common along the coast, growing above the upper foreshore of beaches and rocky headlands throughout W̱SÁNEĆ territory.

Traditional Use: Elsie Claxton and Violet Williams knew of five different names for this wild vetch:

1. DELN̲EW̱ÁLĆEN̲ (*t'əlngəxwél'chəng*), when used as hair tonic; DELN̲EW̱ (*t'élngəxw*) means "medicine".
2. SI,SȻELTEN (*si7skwəl'tən*; probably originally Quw'utsun'), when used in bathing for strength and protection.
3. TATW̱ELTEN (*tátxwəl'tən*; probably originally Quw'utsun'), when referring to its power to protect and strengthen.
4. T̲IQEN-ILĆ (*tl'íkw'ən- ilhch*; meaning "pea/seed"), because it produces peas.
5. SNENW̱Ł-IȽĆ (*snən'xwlh-ílhch*; meaning "canoe-plant"), when used by canoe racers who bathe with it to make them strong.

Giant Vetch grows on the spit at Tsawout, together with Beach Pea. At first Elsie Claxton thought these two plants were the same, but when Violet Williams pointed out that they were different, Elsie agreed.

Elsie said there is lots of Giant Vetch around Michell's farm near Island View Beach. She noted that when you work on it (rub the stems together), the plants are very juicy. She said it was used as a medicine for growing nice long hair. You take the leaves, stems and roots, crush them up and steep them in water. After a few days, you strain the solution, put it in a jar, and then comb it through your hair. It makes your hair shiny and good smelling. Violet's father told her about its use as a hair conditioner. He said that if you use it, you will have beautiful hair, with no split ends, and it will grow right down to your ankles. Christopher Paul also knew

of this hair tonic. He said it was made from the roots of this plant, pounded with a maul and steeped in water, and that it was applied to the hair for falling hair and dandruff.

Violet said that the plant was also used as some kind of love potion. She was told that canoe pullers used the same plant in solution to wash their limbs and rub their bodies after bathing, to keep their muscles fit. It is also used if one has a lot of enemies. The root is rubbed all over your body, and this will protect you and stop your enemies from bothering you anymore.

Elsie's husband's sister, Christine (also Violet's aunt), used to call Giant Vetch TATW̱ELTEN or SNENW̱Ł-IŁĆ. Violet's grandfather called it SI,SȻELTEN, which Violet said refers to something that you pull on and it goes a long way and comes out easily (e.g., when you pull up a root). The name Elsie and Vi used for the plant when it was used as a hair lotion was DELN̲EW̱ÁLĆEN̲. Vi's mother and father used to tell her that the "beach plant with peas", like cedar, was considered to have male and female members.

Cattle Point on San Juan Island is called T̲IQENEN̲ (*tlh'ikw'ən'əng*) "Place of Peas", according to Dave Elliott, probably named after Giant Vetch or Beach Pea.

Other Plants of Cultural Importance to W̱SÁNEĆ People

These plants are arranged in alphabetical order by their scientific names.

Red Columbine   *Aquilegia formosa*
or Sitka Columbine
Violet called this beautiful flower by its Quw'utsun' name, LÁMTEN (*leymtən*),
which means something that guides you or looks after you. Elsie knew of no name
for it, but regarded it as a special plant that would protect you and bring you good
luck.

Common Burdock   *Arctium minus*
Elsie Claxton's mother used to call this plant XEN̲XIN̲ELE, (*x̲əng̲x̲ingələ7*; meaning "chicken hawk"). It is an aggressive introduced weed with large hooked burs
that stick in your hair and clothing.

Kinnikinnick   *Arctostaphylos uva-ursi*
In other Salish areas, people used to toast Kinnikinnick leaves and smoke them
alone or mixed with tobacco. This practice was apparently quite recent for coastal
peoples, having been introduced from the interior around the time of the fur trade.
Elsie recognized this creeping evergreen shrub, but she had not heard of its leaves
being smoked.

Wild Ginger   *Asarum caudatum*
Aromatic wild ginger once formed a patch of dark-green heart-shaped leaves at
Goldstream on the far side of the bridge, but it was destroyed to accommodate
some salmon-spawning channels. It is considered very rare now, and only occurs in a few places, such as in the Westholme and Ladysmith areas. It is called
T̲ET̲LATEN (*t'thət'thle'tən'*; *t'thelə7* means "heart") and people value it as an
important spiritual and good-luck plant.

Red Columbine.

Wild Ginger.

Basket made with Slough Sedge.

Pipsissewa.

Calypso or False Ladyslipper   *Calypso bulbosa*
Violet called this beautiful woodland orchid by its Quw'utsun' name, TI,TEḴELTEN (*ti7təqw'əl'tən'*). Elsie called it SLIPES, a name borrowed from English "slippers". It is a good luck flower.

Slough Sedge   *Carex obnupta*
Apparently, this and other sharp-edged sedges were all called T̲E̲T̲ (*tl'ə'tl'*). Elsie knew that the plant called by this name was sharp to the touch, and that the Nuu-chah-nulth people from the west coast of Vancouver Island used the leaves to make baskets; she said that they dyed the leaves different colours. But neither she nor Violet knew what the plant looked like, and when we saw some Slough Sedge at Goldstream, they did not recognize it as the kind used for making baskets. Large-headed Sedge (*Carex macrocephala*), with large, prickly heads, was appar-ently called by the same name; it grows at T̲IXE̲N̲, the spit at Tsawout. The book *Tsawwassen Legends* (Anon 1961), contains a story about "needlegrass", possibly Large-headed Sedge. In this story, the W̱SÁNEĆ are gathering camas bulbs on one of the Gulf Islands, when enemy canoes suddenly appear. The enemies land and run up the beach expecting an easy victory, when for no apparent reason, they fall to the ground screaming with pain. The W̱SÁNEĆ kill every one of them. It turns out that the enemies ran over a thick patch of needlegrass in their bare feet.

Pipsissewa   *Chimaphila umbellata*
Christopher Paul said that the leaves of this plant were put in the bath water of W̱SÁNEĆ sprinters and canoe-pullers as a liniment for sore muscles. Elsie and Violet did not know this plant.

Pacific Bleeding Heart   *Dicentra formosa*
This fine-leaved, pink-flowered wild flower is considered to be a medicinal plant.

Sweet-scented Bedstraw   *Galium triflorum*
This plant grows at Goldstream. Violet said that it was considered to bring good luck.

Large-leaved Avens   *Geum macrophyllum*
This plant grows at Goldstream and was known as a medicinal plant to Violet Williams. Christopher Paul said that eating the leaves gave one special protective properties when visiting sick or dying people.

Foxtail Barley or Speargrass   *Hordeum jubatum*
This grass, called TAḴEŁ (*téqəlh*), was apparently not used for anything, but was known for its sharp-awned heads that penetrate shoes and clothing.

Common Hop   *Humulus lupulus*
When Violet and Elsie were growing up, their families used to go to Yakima to pick hops along with many others from Vancouver Island. They just called the plant HEPS.

Hairy Cat's-ear   *Hypochaeris radicata*
Violet said that the white, milky juice of this kind of dandelion, like the common dandelion, could be rubbed on warts to get rid of them. She said it was also an eye medicine.

Labrador Tea   *Ledum groenlandicum*
                              syn.: *Rhododendron groenlandicum*
Christopher called this MAḰEM (*méqwəm*) tea or swamp tea, and said that people used to gather the leaves from Rithet's Bog and make tea from them, like other First Peoples up and down the coast. Elsie did not recall having gathered these leaves or using them for tea.

Pineapple-weed   *Matricaria discoidea*
                              syn.: *Matricaria matricarioides*
Elsie said that this small pineapple-scented plant with greenish-yellow chamomile-like flower heads was used as a medicine for coughs and sore throats. People chewed the leaves raw and swallowed the juice. Violet Williams said that when they were kids, they used to eat the little leaves of this plant, but not the yellow flowering heads.

Field Mint or Canada Mint   *Mentha arvensis*
Christopher said that people sometimes used wild mint for flavouring food such as peas. They may have also used the leaves for tea, but this has not been recorded specifically for W̱SÁNEĆ people.

Mock-orange   *Philadelphus lewisii*

People used the straight, hard shoots of this shrub occasionally to make arrow-shafts, according to Christopher. They probably made knitting and mat-making needles from the wood as well.

Pacific Ninebark   *Physocarpus capitatus*

This plant grows in moist swamps and along ditches, and is often seen together with Red-osier Dogwood. Christopher said that people used to macerate the root, steep it in water, and drink the extract as a quick laxative.

Garden Peas   *Pisum sativum*

Peas were called ṬIQEN (*tl'íkwən*) and Elsie Claxton said that there were also wild peas (see Giant Vetch, page 148).

Hooker's Fairybells   *Prosartes hookeri*
           syn.: *Disporum hookeri*

This plant was growing at Goldstream, and both Elsie and Violet said that it looked like a plant they had heard of that was used to bring good luck, but they didn't think it was quite the same.

Domesticated Plum   *Prunus domestica*

These orchard fruits were introduced early in the period of European settlement and were simply called PLEMS.

Pineapple-weed.

Labrador Tea.

American Glasswort or Sea Asparagus — *Salicornia virginica* syn.: *Sarcocornia pacifica*

This succulent plant grows on the tidal flats at Tsawout as well as at Goldstream. Elsie recognized it and said that people in Vancouver eat it and that Chinese people bought it to resell. It was not used by the W̱SÁNEĆ people originally, although Christopher Paul recalled that his father ate a plant of this description when he was fishing at Goldstream.

Hedge-nettle — *Stachys cooleyae*

Christopher said that people used the roots, pounded and steeped in hot water, as a spring tonic: "It really puts the life back into you."

Common Dandelion — *Taraxacum officinale*

This common weedy flower was introduced from Europe around the 1860s and is now widespread in lawns, fields and disturbed ground. When Violet was a child, she rubbed the "milk" (latex) from broken dandelion stems on a wart to get rid of it. She didn't remember how long she kept using it, but one day she looked and the wart was gone. She said that dandelion milk is also good for your eyes when you cannot see very well, but did not elaborate on how it was applied.

Springbank Clover — *Trifolium wormskioldii*

The whitish rhizomes of this native clover were eaten by people all up and down the coast. The W̱SÁNEĆ ate them to some extent, according to Wayne Suttles (1951), digging them with a sharp-pointed stick from some of the islands and steaming them in underground pits.

Western Trillium — *Trillium ovatum*

This beautiful white, three-petalled flower was identified by Violet and Elsie as growing at Goldstream and Coles Bay. The plant was a special medicine for some people.

Springbank Clover.

Appendix 1:
SENĆOŦEN Alphabet
and Equivalent Orthographies

Compiled by Timothy Montler

EO	PO	IPA	EO	PO	IPA
A	e	e	N	n	n / n̓
Á	e	e	N̲	ng	ŋ / ŋ̓
Ⱥ	ey / e	ej	O	a	a
B	p'	p̓	P	p	p
C	k	k	Q	k'w / kw'	k̓ʷ
Ć	ch	ʧ	S	s	s
Ȼ	kw	kʷ	Ś	sh	ʃ
D	t'	t̓	T	t	t
E	ə	ə	T̸	t'th	t͡s̱̓ / tθ̓
H	h	h	T̲	tl'	tɬ̓
I	i	i	Ŧ	th	s̱ / θ
Í	əy / ay	əj / aj	U	u / w	u / əw
J	ch'	ʧ̓	W	w	w / w̓
K	q'	q̓	W̲	xw	xʷ
K̵	qw' / q'w	q̓ʷ	X	x̱	χ
K̲	q	q	X̲	x̱w	χʷ
Ḱ	qw	qʷ	Y	y / y'	j / j̓
L	l	l	Z	ts	ʦ
Ƚ	lh	ɬ	'	'	'
M	m	m / m̓	,	7	ʔ

EO = Elliott Orthography; PO = Practical Orthography;
IPA = International Phonetic Alphabet.

Appendix 2:
Plants with No Recorded Name or Use

Elsie Claxton and Violet Williams did not recognize or have a name for these plants, listed alphabetically by scientific name (in parenthesis).

Sand Verbena (***Abronia latifolia***) grows on the spit at Tsawout – no known name or use.

Thrift or **Sea-pink** (***Armeria maritima***) grows on the spit at Tsawout – no known name or use.

Seaside Wormwood (***Artemisia suksdorfii***) grows on the spit at Tsawout – no known name or use, although Violet commented that its strong smell assured her that it must have been used for something.

Goatsbeard (***Aruncus dioicus***) seen on West Coast Road past Sooke – not known.

Douglas's Aster (***Aster douglasii***) has no known name or use, although both Elsie and Violet had seen it growing. Elsie said, "It's nice, eh?" There is a term, SPAKEN (*speq'əng*), for any kind of flower, including garden flowers and wild flowers. This would be called by that name.

Deer Fern (***Blechnum spicant***) seen by the Sandcut Beach trail, near Jordan River. Violet and Elsie recognized it as a kind of fern, but knew nothing further about it. They called it by the same name as Sword Fern.

Large-headed Sedge or **Big-headed Sedge** (***Carex macrocephala***) grows on the spit at Tsawout (see Slough Sedge, page 151) – no known name or use.

Enchanter's Nightshade (***Circaea alpina***) grows at Goldstream – no known name or use.

 Siberian Miner's-lettuce (***Claytonia sibirica***; syn. *Montia sibirica*) grows at Goldstream – no known name or use.

Scotch Broom (***Cytisus scoparius***) has no known name or use. Violet recognized it and noted that it had yellow flowers. Her mother told her that non-aboriginal people brought it in, and were throwing the seeds all around.

Spiny Wood Fern (***Dryopteris expansa***) seen at Sandcut Beach trail, near Jordan River. Elsie and Violet called it by the same name as Lady Fern.

Fireweed (***Epilobium angustifolium***). Elsie did not seem to recognize a sample of this plant. Christopher Paul said that people used to boil the young leaves for a refreshing tea. The cottony seed fluff was sometimes woven with dog wool and duck feathers to make bags, baskets and mattresses.

Common Stork's-Bill (*Erodium cicutarium*) has no known name or use, though Elsie noticed this plant growing at Tsawout.

Silver Burr Ragweed (*Franseria chamissonis*) grows on the spit at Tsawout – no known name or use.

Chocolate Lily or **Rice-root (*Fritillaria affinis*)**. Violet and Elsie didn't seem to know this plant, but Christopher Paul suggested that the bulbs used to be eaten. Wayne Suttles (1951) documented its former use and recorded its name, *ts'áliqw*. Linguist Geoff O'Grady recorded its name as *stl'əlts'ə'ləwəs* (Turner and Bell 1971).

Bedstraw or **Cleavers (*Galium aparine*)**. Violet's brother told her what it was good for, but she couldn't recall the use or the name.

American Glehnia (*Glehnia littoralis*) grows on the spit at Tsawout – no known name or use.

Entire-leaved Gumweed (*Grindelia stricta*; syn. *Grindelia integrifolia*) grows on the spit at Tsawout – no known name or use.

Seabeach Sandwort (*Honckenya peploides*) seen at Jordan River. Violet picked it up, but neither she nor Elsie recognized it.

Pacific Waterleaf (*Hydrophyllum tenuipes*) grows at Goldstream – no known name or use.

Purple Peavine (*Lathyrus nevadensis*) – not known, but Violet noted that it had peas.

Tiger Lily or **Columbia Lily (*Lilium columbianum*)** – Violet knew this plant, but didn't know its name. She said that it grows at Coles Bay, but she had never heard anything about it from her parents or anyone else. Suttles (1951) recorded the name *ts'ágwit* for the bulbs, and noted that the W̱SÁNEĆ may have eaten them.

Black Twinberry (*Lonicera involucrata*) – not recognized by Violet, Elsie or Mary Thomas.

False Lily-of-the-Valley (*Maianthemum dilatatum*) grows at Goldstream and at the Sandcut Beach trail, near Jordan River – no known name or use.

Wild Cucumber (*Marah oregana*). Violet did not know or recognize this plant, and said that if it grows at Pat Bay (we had a sample from there), it must have been planted there by non-aboriginal people. (J.K. Henry (1915) suggested that it was planted on southern Vancouver Island and the Gulf Islands by indigenous people.)

False Azalea (*Menziesia ferruginea*) seen at Sandcut Beach trail, near Jordan River – not known.

Yellow Pond-lily (*Nuphar polysepala*). Elsie and Violet did not recognize the name, but Violet knew which plant it was – use unknown.

Water-Parsley (*Oenanthe sarmentosa*) – not known.

Sweet Coltsfoot (*Petasites speciosus*; syn. *Petasites frigidus*) grows at Goldstream – no known name or use.

Reed Canarygrass (*Phalaris arundinacea*). Both Violet and Elsie exclaimed that it's a nice, tall grass, but knew of no generic level name or any use for it. They were interested to hear that some mainland basketmakers used the stems to decorate baskets. They had never heard of such a use. They would call it SA̱WEL (*sáxwəl*), the general term for grass and grasslike plants.

Badge Moss (*Plagiomnium insigne*) grows at Goldstream – no known specific name or use, but see Mosses on page 38.

Sea Plantain (*Plantago maritima*) grows on the spit at Tsawout – no known name or use.

Beach Knotweed (*Polygonum paronychia*) grows on the spit at Tsawout – no known name or use.

Pacific Silverweed (*Potentilla egedii*; syn. *Potentilla pacifica*) grows at Goldstream. Violet and Elsie didn't recognize it, even in bloom. Christopher Paul remembered his father digging some kind of roots for food in the Goldstream flats, possibly Pacific Silverweed or Springbank Clover (see page 154), which also used to grow at the Goldstream estuary.

Creeping Buttercup (*Ranunculus repens*) grows at Goldstream – no known name or use.

Little Buttercup (*Ranunculus uncinatus*) grows at Goldstream – no known name or use.

Trailing Black Currant (*Ribes laxiflorum*) seen at Sandcut Beach trail, near Jordan River – not known.

Sheep Sorrel or **Sourgrass (*Rumex acetosella*)** – no known name or use, though both Elsie and Violet seemed to recognize it. Elsie commented that it was like the "coffee" plant (docks – see page 140), but smaller. Christopher Paul noted that children like to eat the sour leaves.

Small-flowered Bulrush or **"Cut-grass" (*Scirpus microcarpus*)** grows at Goldstream, but neither Violet nor Elsie knew of a name for it. When asked about PSX̱WEY (*psxəy*), the name for "Cut-grass" (Turner and Bell 1971), Elsie didn't recognize the word. Violet knew the word, but did not know what plant it refers to; she thought it might be a Quw'utsun' word.

Broad-leaved Stonecrop (*Sedum spathulifolium*) – no known name or use, though both Elsie and Violet recognized this small succulent plant.

Cooley's Hedgenettle (*Stachys chamissonis* var. *cooleyae*) grows at Goldstream – no known name or use. Elsie noted that it had a strong smell.

Tall Fringecup (*Tellima grandiflora*) grows at Goldstream – no known name or use. Violet called it "leaves".

Three-leaved Foamflower (*Tiarella trifoliata*) grows at Goldstream and the Sandcut Beach trail, near Jordan River – no known name or use. (*T. unifoliata* was also at Sandcut, but not known.)

False Bugbane (*Trautvetteria caroliniense*) grows and at Goldstream (in full bloom when we were there) – no known name or use.

Seaside Arrow-grass (*Triglochin maritimum*) seen at Jordan River in a floodplain salt marsh – not known.

Stream Violet or **Yellow Wood Violet (*Viola glabella*)** grows at Goldstream – no known name or use.

Meadow Death-camas (*Zigadenus venenosus*). Neither Elsie nor Violet had ever heard of a poisonous bulb plant similar to the edible blue camas (see page 118, and also the warnings on pages 118 and 124).

Glossary

achene   A small, dry one-seeded fruit that does not open to release the seed.

adze   A tool for cutting away the surface of wood. The arched or angled blade is positioned at a right angle to the shaft, like a garden hoe rather than an axe.

alternate   Refering to leaves or buds that grow at the nodes along a stem, alternating on one side of the stem and then the other and so on, rather than in pairs directly opposite to each other.

annual   A plant that lives for only one year.

aril   An extra seed covering, often fleshy or hairy.

basal   At or emerging from the base of a plant or structure.

bract   A modified leaf, either small and scale-like or large and petal-like.

cambium   A layer of continuously dividing cells between the wood and the bark of a tree or shrub, from which new wood and bark tissues are formed.

catkin   A long, drooping cluster of tiny flowers on willows, alders and birches.

conifer   A cone-bearing tree, such as a pine, fir or spruce.

deciduous   Referring to a tree or shrub that sheds its leaves annually.

fiddlehead   A young, curled fern frond.

frond   The divided or compound leaf of a fern.

grilse   A young salmon that has returned to fresh water from the sea for the first time.

herbaceous   Refering to a plant that is not woody; relating to a herb.

holdfast   A root-like structure that attaches a seaweed to the surface it grows from.

humus   The surface part of soil that is formed by the decomposition of plant material.

introduced   Referring to a plant or animal that arrived in a place outside of its natural range because of human activities, either accidentally or intentionally.

lenticel   A raised pore or ridge on the surface of the bark of certain trees and shrubs.

opposite   Referring to leaves or buds that grow in pairs directly opposite each other on a stem, rather than alternating; see *alternate*.

perennial   A plant that lives for more than two years.

receptacle   A specialized stem tip bearing some or all of the flower parts.

recurving   Bending backwards.

rhizome   A creeping underground stem, often fleshy, that serves in vegetative reproduction and food storage for the plant.

rosette   A circular arrangement of horizontally spreading leaves or bracts, resembling a rose flower, about the base of a stem.

sepal   A division of the calyx, the protective cover of a flower bud. When the flower opens, the sepals fan out just under the petals.

shrub   A relatively small woody perennial, usually with several permanent stems instead of a single trunk, like a that of a tree.

sori   Clusters of spore cases on the undersurface of a fern frond (plural of sorus).

spadix   A spike of flowers closely arranged around a fleshy axis.

stipe   The slender stalk that connects a fern blade to the stem of a frond.

stolon   A horizontal stem or branch that takes root along its length, producing new plants.

taproot   The main root that grows vertically downward, from which smaller branch roots grow out, like a carrot.

umbel   A flower cluster in which stalks of about the same length grow from a common centre and form a flat or slightly curved surface that resembles an umbrella.

warp   In the weaving of baskets, mats, etc., strands of fibre stretched lengthwise to be crossed by the weft.

weft   In the weaving of baskets, mats, etc., strands of fibre woven across the warp.

withe   A tough, flexible twig.

Acknowledgements and Credits

Our heartfelt thanks go to all those who have participated in developing this book, especially to the elders: Elsie Claxton, Dave Elliott Sr, Christopher Paul and Violet Williams. The families of these people are also gratefully acknowledged. In particular, we extend deep appreciation to Dr Earl Claxton Sr and Dr John Elliott, whose help with this book was especially valuable. Both of these men have followed the spirit of their parents' teachings – and have stepped into their shoes as cultural and language leaders and teachers. This book would not have been possible without their help and dedication. Sadly, Dr Earl Claxton Sr passed away in 2011.We express our deep thanks to Marguerite Babcock, M.Ed., who, as a student, worked with Christopher Paul in the late 1960s and recorded detailed information on blue camas, which otherwise would have been lost. She lives in Pennsylvania, and has given permission for her notes and writings to be cited.

Linguistic transcriptions by Dr Earl Claxton Sr (YELḰÁTₜE) and Dr Timothy Montler, based on writing system developed by Dave Elliott Sr, with earlier transcriptions by Timothy Montler, Department of Linguistics and Technical Communication, University of North Texas, Denton, working with Elsie Claxton and Violet Williams.

Illustrations and content advice by Dr John Elliott (STOL₵EŁ) and Linda Elliott, Belinda Claxton (Selliliye), Earl Claxton Jr, and the family of Violet Williams, especially Doris and Maureen.

Others who have helped us over the years in various aspects of the book production include: Belinda Claxton, Earl Claxton Jr, Dan Claxton, John Bradley Williams, Jordan Bennett, Melissa Kingan Grimes, Janet Leonard, Chris Harvey, Robert D. Turner, Glenn Bartley, Brian Seymour and Carlo Mocellin. Dr Tim Montler read over and corrected the transcriptions of Saanich names, in both the SENĆOŦEN and more standard linguistic alphabets. He also prepared the table in Appendix 1, showing the equivalent orthographies for the different sounds represented in SENĆOŦEN, the practical linguistic alphabet and the International Phonetic Alphabet. We are grateful to Gerry Truscott for shepherding this book through to completion in his capacity as RBCM publisher.

About the Authors

Dr Nancy J. Turner is Distinguished Professor and Hakai Professor in Ethno-ecology in the School of Environmental Studies at the University of Victoria. She has published more than 20 books and over 125 articles on ethnobotany and First Nations issues. She has received numerous awards for her work and is a member of the Order of Canada and the Order of British Columbia.

Dr Richard J. Hebda is Curator of Earth History and Botany at the Royal BC Museum and Adjunct Professor in the Department of Biology and School of Earth and Ocean Sciences at the University of Victoria. He has written hundreds of articles and several books on BC's plants and ice-age history.

Credits and Copyright Information

Edited, produced and typeset in Times New Roman (10/12) by Gerry Truscott, with editorial assistance from Alex Van Tol.
Cover design by Jenny McCleery.
Index by Carol Hamill.

References

Adams, R.P. 2007. "Juniperus maritima, the seaside juniper, a new species from Puget Sound, North America." *Phytologia* 89(3): 263-83.

Anon. 1961. *Tsawwassen Legends*. Ladner, BC: The Optimist, Dunning Press.

Arora, David. 1986. *Mushrooms Demystified: A Comprehensive Guide to the Fleshy Fungi*. Berkeley, CA: Ten Speed Press.

Arora, David. 1991. *All That The Rain Promises and More: A Hip Pocket Field Guide*. Berkeley, CA: Ten Speed Press.

Babcock, Marguerite. 1967. "The Gathering and Preparing of Camas in Southern Vancouver Island." Unpublished manuscript, copy with N.J. Turner.

Barnett, Homer G. 1955. *The Coast Salish of British Columbia*. Portland: University of Oregon Press.

Claxton, Earl Sr, and John Elliott Sr. 1993. *The Saanich Year*. Brentwood Bay, BC: Saanich Indian School Board.

Claxton, Earl Sr, and John Elliott Sr. 1994. *Reef Technology of the Saltwater People*. Brentwood Bay, BC: Saanich Indian School Board.

Henry, J.K. 1915. *Flora of Southern British Columbia*. Toronto: W.J. Gage.

Jenness, Diamond. 1945. "The Saanitch Indians of Vancouver Island." Unpublished manuscript. National Museum of Civilization, National Museums of Canada, Ottawa.

Jenness, Diamond. n.d. (ca 1930). "Coast Salish Field Notes." Unpublished manuscript (no. 1103.6) in Ethnology Archives, National Museum of Civilization, Ottawa.

Elliott Sr, David. 1980. Unpublished, untitled manuscript on Saanich plants, in the possession of John Elliott.

Elliott, David, Earl Claxton Sr, Gabriel Bartleman and Linda Underwood, with G.E. Mortimore. 1987. "I Remember the Names of the Winds." Unpublished manuscript, in possession of Earl Claxton and John Elliott.

Meidinger, D.V., and Pojar, J.J. 1991. *Ecosystems of British Columbia.* Special Report Series, 6. Victoria: BC Ministry of Forests.

Suttles, Wayne P. 1951. "Economic Life of the Coast Salish of Haro and Rosario Straits." PhD thesis. University of Washington, Seattle.

Suttles, Wayne P. 1987. *Coast Salish Essays.* Seattle: University of Washington Press.

Suttles, Wayne P., ed. 1990. *Northwest Coast.* Handbook of North American Indians, vol. 7. Washington, DC: Smithsonian Institution.

Thompson, Laurence C., and M. Dale Kinkade. 1990. "Northwest Coast Languages." In *Northwest Coast.* Handbook of North American Indians, vol. 7, edited by Wayne Suttles. Washington, DC: Smithsonian Institution.

Turner, Nancy J. 1995. *Food Plants of Coastal First Peoples.* Royal BC Museum Handbook (2006 edition). Victoria: Royal BC Museum.

Turner, Nancy J. 1998. *Plant Technology of First Peoples in British Columbia.* Royal BC Museum Handbook (2007 edition). Victoria: Royal BC Museum.

Turner, Nancy J., and Marcus A.M. Bell. 1971. "The Ethnobotany of the Coast Salish Indians of Vancouver Island." *Economic Botany* 25 (1): 63-104.

Turner, Nancy J., and Richard J. Hebda. 1990. "Contemporary Use of Bark for Medicine by Two Salishan Native Elders of Southeast Vancouver Island, Canada. *Journal of Ethnopharmacology* 29 (1990): 59-72.

Turner, Nancy J., L.C. Thompson, M.T. Thompson and A.Z. York. 1990. *Thompson Ethnobotany: Knowledge and Usage of Plants by the Thompson Indians of British Columbia.* Victoria: Royal BC Museum.

Index

Index to SENĆOŦEN Language

ROYAL **BC** MUSEUM

British Columbia is a big land with a unique history. As the province's museum and archives, the Royal BC Museum captures British Columbia's story and shares it with the world. It does so by collecting, preserving and interpreting millions of artifacts, specimens and documents of provincial significance, and by producing publications, exhibitions and public programs that bring the past to life in exciting, innovative and personal ways. The Royal BC Museum helps to explain what it means to be British Columbian and to define the role this province plays in the world.

The Royal BC Museum administers a unique cultural precinct in the heart of British Columbia's capital city. This site incorporates the Royal BC Museum (est. 1886), the BC Archives (est. 1894), the Netherlands Centennial Carillon, Helmcken House, St Ann's Schoolhouse and Thunderbird Park, which is home to Wawaditła (Mungo Martin House).

Although its buildings are located in Victoria, the Royal BC Museum has a mandate to serve all citizens of the province, wherever they live. It meets this mandate by: conducting and supporting field research; lending artifacts, specimens and documents to other institutions; publishing books (like this one) about BC's history and environment; producing travelling exhibitions; delivering a variety of services by phone, fax, mail and e-mail; and providing a vast array of information on its website about all of its collections and holdings.

From its inception 125 years ago, the Royal BC Museum has been led by people who care passionately about this province and work to fulfil its mission to preserve and share the story of British Columbia.

Find out more about the Royal BC Museum at www.royalbcmuseum.bc.ca.